Praise for

Half-Earth

An Amazon Best Book of 2016

"Edward O. Wilson possesses a rare, almost unique, combination of immense scientific knowledge and deep humane intelligence. Looking around him at the beloved natural world he has done so much to understand and taking the measure of the massive damage to it caused by human stupidity and greed, he has every reason to succumb to despair. But *Half-Earth* is not a bitter jeremiad. It is a brave expression of hope, a visionary blueprint for saving the planet."

—Stephen Greenblatt, author of *The Swerve*

"If humankind finds a way to live in peace together, and in harmony with nature, Wilson will have played a unique role in that deliverance."

—Jeffrey Sachs

"Wilson's sense of the future cannot be easily dismissed. Over the years, we have learned a lot about the world and ourselves from him; and with *Half-Earth*, a book of vision and welcome optimism, he has yet more to teach us." —Jack E. Davis, *Tampa Bay Times*

"An audacious idea that might jump-start a lagging conversation about a burning issue. . . . [I]f *Half-Earth* takes us any closer to sparking greater effort, it will cement Wilson's already remarkable legacy."

—Mike Weilbacher, *Philadelphia Inquirer*

"Wilson's passion for the planet shines through on these pages. He looks at life in its broadest, grandest sweep. . . . Wilson is a thinker in the tradition of Alexander von Humboldt."

—Matthew Price, *The National*

"Few experts have offered such an exuberant and optimistic plan for dealing with [climate change] as biologist Edward O. Wilson. . . . The strength of his argument lies in his ability to elegantly unveil the bigger picture, and to define and examine what in our essential human nature has led us to this point. . . . [W]e need Wilson's reminder that we are not demigods, but are instead, as he puts it, 'a biological species tied to this particular biological world.'"

—Jessi Phillips, *Sierra*

"As an outline of our terrible ecological plight, it does a first-class job."

—Robin McKie, *Guardian*

"An invigorating elucidation of why the full range of interdependent life forms is necessary for a living Earth, and how habitat destruction, pollution, poaching, overhunting, and overfishing are destroying myriad species. . . . This startling, courageous, many will say wildly quixotic vision of a truly global preservation effort is guaranteed to stoke the fires of environmental debate."

—Donna Seaman, *Booklist*, starred review

Half-Earth

Biological Diversity: The Oldest Human Heritage (1999)

Consilience: The Unity of Knowledge (1998)

In Search of Nature (1996)

Journey to the Ants: A Story of Scientific Exploration,
with Bert Hölldobler (1994)

Naturalist (1994); new edition, 2006

The Diversity of Life (1992)

The Ants, with Bert Hölldobler (1990);
Pulitzer Prize, General Nonfiction, 1991

Success and Dominance in Ecosystems: The Case of the Social Insects (1990)

Biophilia (1984)

Promethean Fire: Reflections on the Origin of the Mind,
with Charles J. Lumsden (1983)

Genes, Mind, and Culture, with Charles J. Lumsden (1981)

On Human Nature (1978); Pulitzer Prize, General Nonfiction, 1979

Caste and Ecology in the Social Insects, with George F. Oster (1978)

Sociobiology: The New Synthesis (1975); new edition, 2000

A Primer of Population Biology, with William H. Bossert (1971)

The Insect Societies (1971)

The Theory of Island Biogeography,
with Robert H. MacArthur (1967)

Bees, flies, and flowers. Alfred Edmund Brehm, 1883–1884.

Half-Earth

OUR PLANET'S FIGHT FOR LIFE

Edward O. Wilson

LIVERIGHT PUBLISHING CORPORATION

A Division of W. W. Norton & Company

Independent Publishers Since 1923

New York | London

For information about permission to reproduce selections from this book, write to
Permissions, Liveright Publishing Corporation,
a division of W. W. Norton & Company, Inc.,
500 Fifth Avenue, New York, NY 10110

For information about special discounts for bulk purchases, please contact
W. W. Norton Special Sales at specialsales@wwnorton.com or 800-233-4830

Manufacturing by Lake Book Manufacturing
Book design by Chris Welch
Production manager: Anna Oler

Library of Congress Cataloging-in-Publication Data

Names: Wilson, Edward O.
Title: Half-earth : our planet's fight for life / Edward O. Wilson.
Description: First edition. | New York : Liveright Publishing Corporation, 2016. |
Includes bibliographical references and index.
Identifiers: LCCN 2015041784 | ISBN 9781631490828 (hardcover)
Subjects: LCSH: Biodiversity conservation. | Biosphere reserves. | Human ecology. |
Nature—Effect of human beings on.
Classification: LCC QH75 .W536 2016 | DDC 333.95—dc23
LC record available at http://lccn.loc.gov/2015041784

ISBN 978-1-63149-252-5 pbk.

Liveright Publishing Corporation,
500 Fifth Avenue, New York, N.Y. 10110
www.wwnorton.com

W. W. Norton & Company Ltd.
15 Carlisle Street, London W1D 3BS

3 4 5 6 7 8 9 0

But now we have come a great long way and now
The time has come to unyoke our steaming horses.

—VIRGIL, *The Second Georgic*
(Translated by David Ferry)

CONTENTS

Prologue 1

PART I. The Problem

PART II. The Real Living World

PART III. The Solution

PROLOGUE

What is man?

Storyteller, mythmaker, and destroyer of the living world. Thinking with a gabble of reason, emotion, and religion. Lucky accident of primate evolution during the late Pleistocene. Mind of the biosphere. Magnificent in imaginative power and exploratory drive, yet yearning to be more master than steward of a declining planet. Born with the capacity to survive and evolve forever, able to render the biosphere eternal also. Yet arrogant, reckless, lethally predisposed to favor self, tribe, and short-term futures. Obsequious to imagined higher beings, contemptuous toward lower forms of life.

For the first time in history a conviction has developed among those who can actually think more than a decade ahead that we are playing a global endgame. Humanity's grasp on the planet is not strong. It is growing weaker. Our population is too large for safety and comfort. Fresh water is growing short, the atmosphere and the seas are increasingly polluted as a result of what has transpired on the land. The climate is changing in ways unfavorable to life, except for microbes, jellyfish, and fungi. For many species it is already fatal.

Because the problems created by humanity are global and

progressive, because the prospect of a point of no return is fast approaching, the problems can't be solved piecemeal. There is just so much water left for fracking, so much rain forest cover available for soybeans and oil palms, so much room left in the atmosphere to store excess carbon.

Meanwhile, we thrash about, appallingly led, with no particular goal in mind other than economic growth, unfettered consumption, good health, and personal happiness. The impact on the rest of the biosphere is everywhere negative, the environment becoming unstable and less pleasant, our long-term future less certain.

I've written *Half-Earth* as the last of a trilogy that describes how our species became the architects and rulers of the Anthropocene epoch, bringing consequences that will affect all of life, both ours and that of the natural world, far into the geological future. In *The Social Conquest of Earth*, I described why advanced social organization has been achieved only rarely in the animal kingdom, and then late in the 3.8-billion-year history of life on Earth. I reviewed the evidence of what transpired when the phenomenon emerged in one species of large-sized African primates.

In *The Meaning of Human Existence*, I reviewed what science tells us about our sensory system (surprisingly weak) and moral reasoning (conflicted and shaky), and why both the system and reasoning are deficient for the purposes of modern humanity. Like it or not, we remain a biological species in a biological world, wondrously well adapted to the peculiar conditions of the planet's former living environment, albeit tragically not this environment or the one we are creating. In body and soul we are children of the Holocene, the epoch that created us, yet far from well adapted to its successor, the Anthropocene.

In *Half-Earth* I propose that only by committing half of the planet's surface to nature can we hope to save the immensity of life-forms that compose it. I'll identify the unique blend of animal instinct and social and cultural genius that has launched our species and the rest of life on a potentially ruinous trajectory. We need a much deeper understanding of ourselves and the rest of life than the humanities and science have yet offered. We would be wise to find our way as quickly as possible out of the fever swamp of dogmatic religious belief and inept philosophical thought through which we still wander. Unless humanity learns a great deal more about global biodiversity and moves quickly to protect it, we will soon lose most of the species composing life on Earth. The Half-Earth proposal offers a first, emergency solution commensurate with the magnitude of the problem: I am convinced that only by setting aside half the planet in reserve, or more, can we save the living part of the environment and achieve the stabilization required for our own survival.*

Why one-half? Why not one-quarter or one-third? Because large plots, whether they already stand or can be created from corridors connecting smaller plots, harbor many more ecosystems and the species composing them at a sustainable level. As reserves grow in size, the diversity of life surviving within them also grows. As reserves are reduced in area, the diversity within them declines to a

---

* I first presented the basic argument for such a globally expanded reserve in *The Future of Life* (2002), and expanded it in A *Window on Eternity: A Biologist's Walk Through Gorongosa National Park* (2014). The term "Half-Earth" was suggested for this concept by Tony Hiss in his 2014 *Smithsonian* article "Can the World Really Set Aside Half the Planet for Wildlife?"

mathematically predictable degree swiftly—often immediately and, for a large fraction, forever. A biogeographic scan of Earth's principal habitats shows that a full representation of its ecosystems and the vast majority of its species can be saved within half the planet's surface. At one-half and above, life on Earth enters the safe zone. Within half, existing calculations from existing ecosystems indicate that more than 80 percent of the species would be stabilized.

There is a second, psychological argument for protecting half of Earth. The current conservation movement has not been able to go the distance because it is a process. It targets the most endangered habitats and species and works forward from there. Knowing that the conservation window is closing fast, it strives to add increasing amounts of protected space, faster and faster, saving as much as time and opportunity will allow.

Half-Earth is different. It is a goal. People understand and prefer goals. They need a victory, not just news that progress is being made. It is human nature to yearn for finality, something achieved by which their anxieties and fears are put to rest. We stay afraid if the enemy is still at the gate, if bankruptcy is still possible, if more cancer tests may yet prove positive. It is further our nature to choose large goals that while difficult are potentially game-changing and universal in benefit. To strive against odds on behalf of all of life would be humanity at its most noble.

PART I

The Problem

The variety of life-forms on Earth remains largely unknown to science. The species discovered and studied well enough to assess, notably the vertebrate animals and flowering plants, are declining in number at an accelerating rate—due almost entirely to human activity.

A medley of fungi. Franciscus van Sterbeeck, 1675.

I

THE WORLD ENDS, TWICE

Sixty-five million years ago, a twelve-kilometer-wide aster-oid, traveling at twenty kilometers a second, slammed into the present-day Chicxulub coast of Yucatán. It blew out a hole ten kilometers deep and one hundred eighty kilometers wide, and rang the planet like a bell. There followed volcanic eruptions, earthquakes, acid rains, and a mountainous ocean wave that traveled around the world. Soot shaded the skies, blocking sunlight and photosynthesis. The darkness held on long enough to finish off most of the surviving vegetation. In the killing twilight the temperature plummeted and a volcanic winter gripped the planet. Seventy percent of all species disappeared, including the last of the dinosaurs. On a smaller scale, microbes, fungi, and carrion flies, master scavengers of the living world, prospered for a time on dead vegetation and animal corpses. But soon they, too, declined.

Thus ended the Mesozoic Era, the Age of Reptiles, and began the Cenozoic Era, the Age of Mammals. We are the culminating and potentially final product of the Cenozoic.

Geologists divide the Cenozoic Era into seven epochs, each defined by its combination of distinctive environments and the kinds of plants and animals living in them. First in time was the Paleocene Epoch, an interval of ten million years during which the diversity of life rebounded through evolution from the end-of-Mesozoic catastrophe. Then came in succession the Eocene, Oligocene, Miocene, and Pliocene epochs. The sixth epoch in the progression was the Pleistocene, a time of advancing and retreating continental glaciers.

The final epoch, formally recognized by geologists and the one in which we live, is the Holocene. Begun 11,700 years ago, when the latest of the continental glaciers began to retreat, it brought a milder climate and what may have been briefly the highest peak in numbers of species in the history of life.

The dawn of the Holocene also found humanity newly settled throughout almost all of Earth's habitable land. All three of the levels in which life is organized faced a new threat with the potential destructive power of the Chicxulub strike. The levels were and remain first the ecosystems, which include coral reefs, rivers, and woodlands; then the species, such as the corals, fishes, and oak trees that make up the living part of the ecosystems; and finally the genes that prescribe the traits of each of the species.

Extinction events are not especially rare in geological time. They have occurred in randomly varying magnitude throughout the history of life. Those that are truly apocalyptic, however, have occurred at only about hundred-million-year intervals. There have been five such peaks of destruction of which we have record, the latest being Chicxulub. Earth required roughly ten million years to recover from each. This is the reason that the peak of destruction that humanity has initiated is often called the Sixth Extinction.

Many authors have suggested that Earth is already different enough to recognize the end of the Holocene and replace it with a new geological epoch. The favored name, coined by the aquatic biologist Eugene F. Stoermer in the early 1980s and popularized by the atmospheric chemist Paul Crutzen in 2000, is the Anthropocene, the Epoch of Man.

The logic for distinguishing the Anthropocene is sound. It can be clarified by the following thought experiment. Suppose that in the far-distant future geologists were to dig through Earth's crusted deposits to the strata spanning the past thousand years of our time. There they would encounter sharply defined layers of chemically altered soil. They would recognize the physical and chemical signatures of rapid climate changes. They would uncover abundant fossil remains of domesticated plants and animals that had replaced, suddenly and globally, most of Earth's prehuman fauna and flora. They would excavate fragments of machines, and a veritable museum of deadly weapons.

"The Anthropocene," far-distant geologists might say, "unfortunately married swift technological progress with the worst of human nature. What a terrible time it was for people, and for the rest of life."

The edge of a European woodland. Alfred Edmund Brehm, 1883–1884.

2

HUMANITY NEEDS A BIOSPHERE

The biosphere is the collectivity of all the organisms on the planet at any given moment in time. It is all the plants, animals, algae, fungi, and microbes alive as you read this sentence.

The upper boundary of the biosphere consists of bacteria swept upward by storms to ten thousand meters and possibly higher. Comprising 20 percent of the microscopic particles found at this altitude (the remainder are inert dust particles), some of the bacterial species are believed to recycle materials and reproduce by photosynthesis and scavenging dead organic matter. Can this high-drifting stratum be called an ecosystem? The matter is still under discussion.

The lower boundary of life exists along the lower edge of what scientists call the deep biosphere. There, at more than three kilometers below the surface on land and sea, bacteria and nematodes (roundworms) survive the intense heat coming up from Earth's magma. The very few resident species found by scientists in this hellish stratum live on energy and materials drawn from rocks around them.

The biosphere, compared with the immense bulk of the planet as a whole, is razor-thin and negligible in weight. Plastered on the surface like a membrane, it cannot be seen sideways with unaided vision from a vehicle orbiting outside Earth's atmosphere.

Deeming ourselves rulers of the biosphere and its supreme achievement, we believe ourselves entitled to do anything to the rest of life we wish. Here on Earth our name is *Power*. God's mocking challenge to Job no longer daunts us.

> Hast thou entered into the springs of the sea? or hast thou walked
>   in the search of the depth?
> Have the gates of death been opened unto thee? or hast thou seen
>   the doors of the shadow of death?
> Hast thou perceived the breadth of the earth? declare if thou
>   knowest it all.
> Where is the way where light dwelleth? and as for darkness, where
>   is the place thereof . . . ?
> . . . Who hath divided a watercourse for the overflowing of waters,
>   or a way of the lightning of thunder . . . ?*

Well, granted we've done all that, more or less. Explorers have descended to the Mariana Trench, and there, in the deepest part of the ocean, they've seen fish and collected microbes. They've even traveled completely away from the planet, although drawing no closer to a now-silent God. Our scientists and engineers have

---

* *Holy Bible*, King James version, Job 38:16–19, 25.

launched vehicles and robots able to examine other planets in the solar system, and asteroids passing by, in minute detail. Soon we'll have the ability to reach other star systems, and the planets that circle them.

Yet we ourselves, our physical bodies, have stayed as vulnerable as when we evolved millions of years ago. We remain organisms absolutely dependent on other organisms. People can live unaided by our artifacts only in bits and slivers of the biosphere, and even there we are severely constrained.

There can be no exception to our extreme flesh-bound fragility. We obey the Rule of Threes used by the military and others in survival training: *You can live for three minutes without air, three hours without shelter or proper clothing in freezing cold, three days without water, and three weeks without food.*

Why must human beings be so weak and dependent? For the same reason all the other species in the biosphere are comparably weak and dependent. Even tigers and whales require protection in a particular ecosystem. Each is delicate in its own way, each is constrained by its own version of the Rule of Threes. To make the point, if you increase the acidity of a lake, certain species in it will disappear, but not all. Some of the survivors, having relied on the presence of the newly extinct species—mostly as providers of food and for protection against predators—will also in time disappear. The population effect of this kind of interaction, called by scientists density-dependent regulation, is a universal rule of all life.

A textbook example of density-dependent regulation is the role wolves have played in the promotion of tree growth. In Yellowstone National Park, the presence of no more than a small pack of wolves in the vicinity drastically reduces the number of elk in the same area. One wolf can consume most of the body of an elk in a week (it can digest a full meal in hours), while one elk can literally mow down a large number of aspen seedlings in the same time. Even the mere presence of the canine top-level predator is enough to frighten elk from the neighborhood. When wolves are present, fewer aspen are browsed by elk, and the aspen groves thicken. When the wolves are removed, the elk return and the growth of aspen plummets.

In the mangrove forest of the Sundarbans National Park of India and the Sundarbans Reserve Forest of Bangladesh, tigers play the same role, preying on and thinning the populations of spotted deer, wild boar, and macaques (and humans, unfortunately), promoting a lusher, more biologically diverse fauna and flora.

Biodiversity as a whole forms a shield protecting each of the species that together compose it, ourselves included. What will happen if, in addition to the species already extinguished by human activity, say, 10 percent of those remaining are taken away? Or 50 percent? Or 90 percent? As more and more species vanish or drop to near extinction, the rate of extinction of the survivors accelerates. In some cases the effect is felt almost immediately. When a century ago the American chestnut, once a dominant tree over much of eastern North America, was reduced to near extinction by an Asian fungal blight, seven moth species whose caterpillars depended on its vegetation vanished, and the last of the passenger pigeons plunged to extinction. As extinction mounts, biodiversity reaches a tipping point at which the ecosystem collapses. Scientists have only begun to study under what conditions and when this catastrophe is most likely to occur.

In one realistic disaster scenario, a habitat can suffer a complete take-over by alien species. This event is not a Hollywood script. In every country where biodiversity censuses are kept, the number of colonist species is rising exponentially. Among those, a few to some degree harm humans, the environment, or both. By presidential executive order in the United States, to help clarify government policy they are designated "invasive" species. A small percentage of invasive species cause major damage, with the potential of reaching catastrophic levels. They include species whose destructiveness has made them household names. Their swiftly growing roster includes the imported fire ant, Asian termite ("the termite that ate New Orleans"), gypsy moth, emerald elm beetle, zebra mussel, Asian carp, snakehead, two species of python, and the West Nile virus.

The invasives come from parts of the world where they have lived for millennia as native species. Because they are naturally adapted with other native species in their homeland, variously in the roles of predators, prey, and competitors, their populations as a whole are kept under control. In their homeland, as it also turns out, invasives tend to be adapted for life variously in grasslands, on riverbanks, and in other habitats of the kind favored by humans. The imported fire ant, scourge of the American South with its hot-needle sting, does best as an invasive in pastures, yards of residences, and road banks. In its South American native range it is mostly a well-behaved specialist of grasslands and floodplains. (A note of caution about this species: Imported fire ants have been a favorite subject of mine for field and laboratory research. I once put my hand briefly into a nest as part of a filmed demonstration, and within seconds received fifty-four stings from the enraged worker ants. Within twenty-four hours each sting turned into an itching pustule. My advice: never put your hand into, much less sit on, fire ant nests.)

Other invasives do not live in human habitat but can be particularly dangerous to the natural environment. The little fire ant, a species smaller than the common fire ant (and another subject of my research), is a native of the South American rain forest. It is able to penetrate tropical forests elsewhere in dense swarms, where it proceeds to decimate single-tarsused (insect term for single-handed) almost all of the other invertebrates inhabiting the leaf litter and soil.

Another, horrific habitat killer is the brown tree snake, which was introduced by accident onto the island of Guam in the late 1940s from New Guinea or the Solomon Islands. Being mostly specialized to feed on nesting birds, it wiped out every songbird of several species on Guam, down to the last individual.

It is contrary to all evidence to suppose, as a few writers have, that in time invasive species will settle down with native species into stable "new ecosystems." Quite the contrary. The only proven way to halt the destabilization of the living world is to protect the largest possible reserves and the native biodiversity surviving within them.

Human beings are not exempt from the iron law of species interdependency. We were not inserted as ready-made invasives into an edenic world. Nor were we intended by providence to rule that world. The biosphere does not belong to us; we belong to it. The organisms that surround us in such beautiful profusion are the product of 3.8 billion years of evolution by natural selection. We are one of its present-day products, having arrived as a fortunate species of Old World primates. And it happened only a geological eyeblink ago. Our physiology and our minds are adapted for life in the biosphere, which we have only begun to understand. We are now able to protect the rest of life, but instead we remain recklessly prone to destroy and replace a large part of it.

Life cycle of a moth (larva, pupa, winged adult) on the food plant of the larva (caterpillar). Maria Sibylla Merian, 1679–1683.

3

HOW MUCH BIODIVERSITY
SURVIVES TODAY?

The total number of species alive on Earth can in theory be counted. In time we will be able to write the number down within narrow limits. But for the moment, conservation scientists look at a world census as a dilemma wrapped in a paradox. The amount of Earth's biodiversity, we have found, is like a magic well. The more species humanity extinguishes, the more new ones are discovered. But this only adds to any estimate of the magnitude of destruction under way in species per year. We need to apply the approximate extinction rate of known species also to those that are unknown. There is no reason as yet to suppose that the two groups of species, known and still unknown, differ from each other radically. That realization leads to a dilemma that turns out also to be one of the great moral questions of all time: Will we continue to degrade the planet to satisfy our own immediate needs, or will we find a way to halt the mass extinction for the sake of future generations?

If we choose the path of destruction, the planet will continue to descend irreversibly into the Anthropocene Epoch, the biologically final age in which the planet exists almost exclusively by, for, and of ourselves. I prefer to call this option by another name, the Eremocene, the Age of Loneliness. The Eremocene is basically the age of people, our domesticated plants and animals, and our croplands all around the world as far as the eye can see.

To measure the biosphere and its rate of diminution, the best unit to use by far is the species. Individual ecosystems, which are composed of species, are much more subjective in their boundaries. One thinks of foothill scrubland grading into mountain forests, oxbow lakes into rivers, riverbanks into deltas, and ground-soaking seeps into springs. Genes, which prescribe the defining traits of species, are on the other hand objective and can be exactly defined, but are more difficult to read and use to meet the multiple needs of taxonomy and biology. You can with binoculars census a medley of warblers as they fly from one ecosystem to another—say, forest edge to forest interior—but identifying their preferred habitat is difficult, and sequencing their DNA for identification is even more so without capturing or killing specimens.

Far more important, however, is the circumstance that the traits with which we recognize organisms are the ones that they themselves use, variously by sight, sound, and smell. With species in mind we are able to understand how life evolves, and how and why each life-form is unique in its combination of anatomy, physiology, behavior, habitat preference, and every other property by which it survives and reproduces.

Biologists define species as populations of individuals that mostly share the same traits and in addition interbreed freely among themselves under natural conditions, but not with other species. The textbook case of proven species is provided by the lion and tiger. The two big cats will interbreed when caged together—but not in nature. In ancient times their geographic ranges overlapped across a broad region, lions through all of Africa, including the Mediterranean coast, then east to India (where a small population still survives in Gujarat), and tigers from the Caucasus to easternmost Siberia. No hybrids have ever been reported among wild populations, in either ancient or recent centuries.

In 1758 Carl Linnaeus, a professor of botany at the University of Uppsala, published the system of classification that biologists have used to the present time. The goal he set was to describe all of the species of plants and animals in the world. With the help of his students, who traveled as far away as South America and Japan, Linnaeus accounted for about twenty thousand species. By 2009, according to the Australian Biological Resources Study, the number had grown to 1.9 million. Since then, new species have been discovered and provided formal, Latinized double names (for example, *Canis lupus* for the wolf) at the rate of about eighteen thousand a year. Thus in 2015 the number of species known to science passed two million.

That figure, however, is still far short of the actual number of living species. Earth, all experts agree, remains a little-known planet. Scientists and the public are reasonably familiar with the vertebrates (fishes, amphibians, reptiles, birds, mammals), mostly because of their large size and immediate visible impact on human life. The best known of the vertebrates are the mammals, with about fifty-five hundred species known and, according to experts, a few dozen remaining to be discovered. Birds have ten thousand recognized species, with an average two or three new species turning up each year. Reptiles are reasonably well known, with slightly more than nine thousand species recognized and a thousand estimated to await discovery. Fishes have thirty-two thousand known and perhaps ten thousand awaiting discovery. Amphibians (frogs, salamanders, wormlike caecilians), among the most vulnerable to destruction, are surprisingly less well known than the other land vertebrates: a bit over sixty-six hundred species discovered and a surprising fifteen thousand believed to exist. Flowering plants come in with about two hundred seventy thousand species known and as many as eighty thousand awaiting discovery.

For most of the rest of the living world, the picture is radically different. When expert estimates for invertebrates (such as the insects, crustaceans, and earthworms) are added to algae, fungi, mosses, and other lower plants; to gymnosperms, as well as flowering plants, bacteria, and other microorganisms, the total added up and then projected has varied wildly, from five million to over one hundred million species.

In 2011, Boris Worm and his fellow researchers at Dalhousie University devised a new way to estimate the number of species, both known and still undiscovered. They proposed to scale downward across the taxonomic categories, ending up at species. First, the number of all phyla (such as mollusks and echinoderms) in the animal kingdom was plotted, then the number of classes in all these phyla, followed by order, family, genus—and finally species. The numbers from phyla to genus are relatively stable, with each rising in a smoothly downwardly bending curve as more and more time is devoted to them. If the shape of these curves is then extended to species, the number of species in the animal kingdom predicted to exist on Earth settles at a quite reasonable 7.7 million. The total number in the Eukarya, which includes plants, animals, algae, fungi, and many kinds of eukaryotic microorganisms (those with mitochondria and other organelles), comes to approximately 8.7 million, give or take a million.

The Dalhousie method might undershoot the mark, however. Many species remain undiscovered for a reason well understood by field biologists. These biologists have learned that the most elusive species tend to be rare and isolated in narrow niches limited to small, remote habitats, and hence could be much more numerous than suggested by published data sets.

Wherever the censuses of biodiversity come to rest among the scientists, the total will be strikingly higher than the two million species so far discovered, given a Latinized double name, and counted to the present time. It is entirely possible that specialists have discovered only 20 percent, or fewer, of Earth's biodiversity at the species level. Scientists working on biodiversity are in a race to find as many of the surviving species as possible in each assemblage—mammals and birds, to tardigrades and tunicates and lichens and lithobiid spiders and ants and nematodes—before they vanish and thus are not only overlooked but never to be known.

Most people are unaware that this unfinished mission of science to discover and conserve all of life on Earth even exists. They have grown accustomed to media accounts that trivialize the subject with headlines such as "Three new kinds of frogs discovered in Mexico" and "Himalayan bird found to consist of two species." Readers are led to believe that the exploration of the living world is nearly finished, so that the discovery of new species is a notable event. As a curator of insects at Harvard University's Museum of Comparative Zoology for much of my career, I can testify how misleading and stultifying that image can be. The truth is that new species flood museums and laboratories everywhere, all the time. Specimens of the novelties pile up in most groups of organisms. They often must wait for years or even decades before the curators of the world's hugely understaffed museums can get to them. The knowledge their study might give us of biology could be put off indefinitely.

If the current rate of basic descriptions and analyses continues, we will not, as I and others have often pointed out, complete the global census of biodiversity—what is left of it—until well into the twenty-third century. Further, if the conservation of Earth's fauna and flora is not more expertly mapped and protected, and soon, the amount of biodiversity will be vastly diminished by the end of the present century. Humanity is losing the race between the scientific study of global biodiversity and the obliteration of countless still-unknown species.

I can illustrate the taxonomic overload very well with an example from my own experience. A part of my research on ants has been on classification, which in any part of biodiversity studies is an absolutely necessary prelude for work on ecology and evolution. Over the years I've described about 450 new species of ants. Of these, 354 were in the single genus *Pheidole*. (As a reminder, a genus is a cluster of species similar to one another and all evolved from the same ancestral species. For example, *Homo* is our genus, and the ancestral species includes *Homo sapiens* and our direct ancestral species, first *Homo habilis*, followed by *Homo erectus*.)

Pheidole, whose name from the Greek means "thrifty one," is the largest and most diverse genus among all of the fourteen thousand known living species of ants. One of the species I discovered and named is *Pheidole scalaris*, where *scalaris* means "ladder," referring to the distinctive ladderlike sculpture on the head of the soldier caste.

Another is *Pheidole hasticeps*, "spearhead," for the spear-headed shape of the soldier's head, and a third is *Pheidole tachygaliae*, "belongs to tachigalia," for the tree (*Tachigalia*) in which the ants make their nest. *Pheidole aloyai* honors Dr. D. P. Aloya, the Cuban entomologist who collected the first specimens in the field. With hundreds of species of *Pheidole* named by me and earlier taxonomists in this fashion, I was running out of Greek and Latin words to describe additional new species. It helped to use the names of collectors like Aloya, and of the localities where the specimens had been found. Then I thought of another way to ease the difficulty. I asked Peter Seligmann, president of Conservation International, to recommend eight members of the CI board of directors who had been distinguished by their private efforts on behalf of global conservation. One of those selected, a fellow board member and friend of mine, now has his own personal ant: *Pheidole harrisonfordi*. There is also a *Pheidole seligmanni*.

Scientific naturalists, amateur and professional both, become acquainted with the species they study almost as though they were other people. One of my mentors when I was an undergraduate at the University of Alabama, the lepidopterist Ralph L. Chermock, once remarked to his students that a true naturalist knows the names of ten thousand species of organisms. I've never come close to that number, and I doubt that Chermock ever did, either. Perhaps a mnemonist could accomplish the feat from illustrations and museum specimens, but he would have little feel or substance in such knowledge alone. Chermock and I could do something better, however. Among the several hundreds of species we had individually studied well, we knew not just the names but the higher categories to which they belong, from phyla through orders to families.

We also knew a great many genera (plural of genus) that especially interested us. We could then further identify at a higher category thousands of species placed in front of us. And more than any but the most dedicated mnemonists, we could add facts and impressions about the biology of the specimens. There would be major gaps, of course, but we could say something or other of use, such as, "That's a *Demognathus* salamander or close to it. I've seen several species. Very common. They prefer terrestrial but very wet habitats; there are several species in the southeastern United States." Or, "That's a solifugid; they're called sun spiders; some people call them camel spiders; they look a little like spiders but they're very different in a lot of ways. They are fast, and I believe all are predators; you find them in the deserts in the Southwest and all over Africa; I've seen a couple of species." Or, "Now, there's something you don't see every day. It's a terrestrial planarian, a flatworm. This is only the second one I've personally ever seen. Most are aquatic or marine, but this one is terrestrial; I believe it occurs all around the world, probably carried accidentally in cargo."

A great majority of people have little awareness of the countless species of the great biosphere that still envelops our planet. In particular, common knowledge of the world-dominant invertebrates, the little things that run the natural world, has dwindled to almost nothing. The working vocabulary of the average person comprises "cockroaches, mosquitoes, ants, wasps, termites, butterflies, moths, bedbugs, ticks, crabs, shrimp, lobsters, earthworms," and a few others consisting of one to several species that, more importantly, affect them personally. The millions of species that support the living world and ultimately our own survival have been reduced to "critters" and "bugs." Within this black night of ignorance we have suffered a massive failure of education and media attention.

The average persons, with busy lives of their own, can't be expected to know Latin and Greek or summon the official two-part names of species. But it would bring a new warmth and richness into their own lives to understand the majesty of biodiversity, even the little bit of it to be seen in no more than a search around their homes. Dedicated naturalists will tell you what it is like to experience twenty kinds of warblers in the migratory season, a dozen species of hawks, or every kind of local mammal save Big Foot.

Pick for a final example any kind of butterfly at random. The thrill of my life as a very young butterfly collector was my first great purple hairstreak, a flying gem not easy to find. I didn't know that its caterpillars feed on the leaves of the mistletoe, a parasitic shrub that grows high in the canopy of trees. I later came to realize that hairstreaks as a whole are the warblers of the butterfly world. Bright in color, varying widely in their geographic range, their habitat, the plant food on which they depend, their abundance or rarity, here are (for example) the common names of the twenty-two species found in the North American East Coast: Acadian, amethyst, banded, Bartram's scrub, coral, early, Edwards', fulvous, gray, great purple, Hessel's, hickory, juniper, King's, mallow scrub, martial scrub, oak, red-banded, ruddy, silver-banded, striped, and white. (Each, of course, also bears a Latinized two-part scientific name.)

Each species is a wonder to behold, a long, brilliant history in itself to read, a champion emerged in our time after a long struggle of thousands or millions of years, best of the best, an expert specialist in the niche of the natural environment in which it lives.

Great Indian rhinoceroses. Alfred Edmund Brehm, 1883–1884.

4

AN ELEGY FOR THE RHINOS

Twenty-seven thousand rhinoceroses remain in the world. A century ago millions thundered across the African plains or slipped silently through the Asian rain forests. They represent five species, all endangered. A large majority of the survivors belong to the southern race of the white rhino, found mostly in South Africa, where they are protected closely by armed guards.

On October 17, 2014, Suni, one of the last surviving northern white rhinos, died in the Ol Pejeta Conservancy of Kenya. His death brought the number of living northern whites in the world down to six: three at Ol Pejeta, one in the Dvůr Králové Zoo in the Czech Republic, and two in the San Diego Zoo Safari Park. The animals are aging, and no young are being produced. With the last of their kind scattered around the world, and rhinos generally difficult to breed in captivity, the northern white rhino is functionally extinct. Taking natural longevity into consideration, it seems certain that the last individual will have died by 2040.

Meanwhile, the western race of the black rhino is totally extinct—

no individuals left anywhere, not even in captivity. Once these great animals with the long curving horn were a symbol of African wildlife. They teemed throughout savannas and dry tropical forests from Cameroon to Chad, then south to the Central African Republic and northeast to the Sudan. The whittling away of their numbers began first with sporting hunters of the colonial era. Then came poachers harvesting rhino horn to make the hafts of ceremonial daggers, principally in Yemen but also in other parts of the Middle East and in North Africa. Finally came the crushing blow, the huge appetite in China and Vietnam for powdered rhino horn as a pharmaceutical of traditional Chinese medicine. The increased consumption was fueled through the favoring by Mao Zedong of traditional Chinese medicine over Western medicine. It is still used for a wide range of ailments, including sexual disorders and cancer. China's population had risen by 2015 to 1.4 billion. So even though only a tiny percent sought rhino horn, the effect on rhinos turned catastrophic. The price per gram has soared to that of gold. The result is a bitter irony: rhinos are being driven to extinction even though their horn has no more medicinal value than a human fingernail.

The rhino horn market has summoned an array of poachers and criminal gangs willing to hunt down the last animal, risking their lives for a dead object you can hold in both hands. There seems to be no stopping the impact on all five rhino species. The western race of the black rhino population declined by 98 percent during 1960–1995. Cameroon, its last stronghold, harbored only fifty individuals in 1991, then thirty-five in 1992. The poacher scourge stayed unrelenting, and the Cameroon government could offer no solution. By 1997, only ten rhinos were left. Unlike white rhinos, which are prone to forming herds of up to fourteen individuals (herds of rhinos are called "crashes," incidentally), black rhinos are solitary except when breeding. In the last days of the western black rhinos, the survivors were scattered across a large area of northern Cameroon. Only four were close enough to one another to meet and mate. They did not follow through, however, and soon all were gone. Millions of years of evolutionary glory came to an end.

The rarest large land mammal in the world at this time is the Javan rhino. A dweller of deep rain forest, the species originally ranged from Thailand to southern China, thence to Indonesia and Bangladesh. Until recently ten Javan rhinos remained hidden and mostly unnoticed in an unprotected forest of northern Vietnam, now the Cát Tiên National Park. Soon after their presence became more widely known, all were killed by poachers. The last was shot in April 2010.

Today the last surviving population is sheltered in the Ujung Kulon National Park, at the westernmost tip of Java. It consists of fewer than fifty individuals. (I have been given a figure of thirty-five by one expert.) A tsunami or determined band of poachers can take the species out in a single strike.

Comparable in both rarity and peril is the Sumatran rhino, another species of tropical Asia's deep rain forest. Once Sumatrans ranged with Javan rhinos widely through southeastern Asia. With much of its habitat replaced by agriculture, and its populations driven downward by the relentless poachers, the species is now limited almost entirely to a few captives in zoos and the dwindling forests of Sumatra. Several more individuals may possibly be hiding in a remote corner of Borneo.

From 1990 to 2015, the worldwide population of the Sumatran rhino plunged to three hundred, then one hundred. In a heroic effort, the veterinarian Terri Roth and her team at the Cincinnati Zoo and Botanical Garden learned how to apply modern human reproductive technology to rhinos. They were successful: three generations have now been produced, allowing a very cautious return of several pioneer animals to reserves in Sumatra. The process is slow, difficult, expensive, and far from assured of success. There are always the sleepless poachers, each willing to put his life on the line for one horn and the lifetime income it will bring.

If the efforts of the captors and Indonesian park guards fail and the Sumatran rhino disappears, it will end an extraordinary line of great animals that has persisted, slowly evolving, for tens of millions of years. Its closest relative, the woolly rhinoceros of the arctic Northern Hemisphere, vanished during the last Ice Age. It was likely driven to extinction by hunters, who (in Europe at least) sketched drawings of it on cave walls for their own and now our delectation.

In late September 1991, during a visit to the Cincinnati Zoo, I was invited by the Director, Ed Maruska, to see a pair of Sumatran rhinos that had been newly captured and transferred from Sumatra by way of the Los Angeles Zoo. One, named Emi, was a female. The other, Ipuh, was a male. Both were young and healthy, but not for long. Sumatran rhinos live only about as long as domestic dogs.

In the early evening we entered an empty warehouse near the zoo. Loud, weirdly irrelevant rock music pounded the interior walls. Maruska explained that the noise was for the rhinos' protection. Occasionally airliners passed close overhead back and forth to the Cincinnati Airport nearby, and at unexpected times they were joined by the sirens of police cars and fire trucks passing in the adjacent streets. The abrupt noises in the dead of night would have startled the rhinos, possibly causing them to panic, bolt, and injure themselves. Better a facilitation to rock music than a violent response to the equivalent sudden sound of a falling tree or approach of tigers, the true dangers of their homeland, or the footfall of hunters—Sumatran rhinos have been exposed to both primitive and modern hunters in Asia for more than sixty thousand years.

That night Emi and Ipuh stood still as statues in their oversized cages. They may have been asleep; I couldn't tell. Drawing close, I asked Maruska if I could touch them. He nodded, and I did so, once each, quickly and softly, with the tips of my fingers. My feeling at that moment was spiritual and lasting, one I can't explain in words to you or even today to myself.

Green sea turtles and men. Alfred Edmund Brehm, 1883–1884.

5

APOCALYPSES NOW

We encountered tropical forests practically devoid of amphibians that were once teeming with dozens of them. We watched mass die-offs. We tried to save threatened species by airlifting them out of infected areas, breeding them in captivity and searching for answers through field and lab research. None of it worked. No cure exists for wild populations. Loss of amphibians continues around the globe. We've seen no significant recovery of populations. Worse, the fungus persists in the environment, preventing the reintroduction of captive animals.

Thus the field biologists Karen R. Lips and Joseph R. Mendelson II have described the devastation of frogs caused by the deadly chytrid fungus, known by the appropriately forbidding scientific name *Batrachochytrium dendrobatidis*. The species has been spread worldwide in freshwater aquariums used to transport frogs, a few of which have been unwittingly infected. And here is a stroke of bad luck: carriers include the African clawed frog

Xenopus, often used in biological and medical research. Then added on is the second, fatal piece of bad luck: the fungus feeds on the entire skins of the adult frogs. Since adults breathe through their skin, they die from suffocation and heart failure.

And if that were not enough, a second chytrid has recently appeared on the scene. Where *Batrachochytrium dendrobatidis* kills frogs, its cousin *Batrachochytrium salamandrivorans* attacks salamanders, the second major group of amphibians. (The second Latin name means "eater of salamanders.") Having invaded Europe from Asia as a free-rider of the pet trade, the parasite inflicts a 98 percent mortality. It looms as an especially dangerous threat to the rich salamander fauna of both temperate and tropical America.

The chytrid invasion for amphibians—frogs and salamanders— is the equivalent of the Black Death for humans that swept Europe in the fourteenth century. In both catastrophes, the descending darkness became a Darwinian tragedy. The predators, having invaded a new continent, found a rich food supply. Their populations exploded, consumed too many of the prey, and doomed themselves to inevitable decline. Humanity has failed the amphibians, especially, to date, the frogs. We should somehow have anticipated and stopped this cruel epizootic.

Frogs and salamanders are important predators that help stabilize moist forests, streamsides, and freshwater wetlands. They are our most gentle neighbors among the vertebrates, the equivalents of birds that we find instead in mud and on leaves of shrubs and in forest litter, beautiful in form and often dazzling in color, timid in demeanor. Frogs sing in choruses during the mating season— sometimes as many as twenty species together in the American tropics—each species with its own song. The whole ensemble at first seems chaotic, but you can learn to distinguish one from another with your eyes closed, by their precise and different scores, as you do with the instruments of an orchestra. During the rest of the year, individuals spread out. Then, if they sing at all, it is a different sound, hard to locate, designed during evolution to mark off territories among others of the same species.

Frogs are pitifully vulnerable. When wetlands and forests are disturbed, they are among the first to disappear. Many of their species are specialized to live only in particular habitats—freshwater marshes or, variously, waterfalls, rock faces, forest canopies, and alpine meadows. And now, as scientists have discovered almost too late, there exist introduced diseases that can erase almost the entirety in one sweep.

I cannot stress enough the menace of invasive species. Some authors, thankfully few in number, have naively suggested that in time alien plants and animals will form "novel ecosystems" that replace the natural ecosystems wiped out by us and our hitchhiker companion species. There is evidence that some alien species of plants "naturalize" in island environments, in other words genetically adapt to them by natural selection. But this occurs only where the diversity of plant species is low and offers a relative abundance of empty niches for aliens to fill.

Allowing the entry of alien species of any kind is the ecological equivalent of Russian roulette. How many cylinders spin in the barrel of the extinction gun? And what percentage of the cylinders are loaded? The answers depend on the identity of the travelers and the niches in the host country they are able to fill. The plants of Europe and North America fit roughly the "tens rule" of invasion biology: by and large, one in ten imported species escapes to the wild, and one in ten of those colonists multiplies and spreads enough to become a pest. For vertebrates (mammals, birds, reptiles, amphibians, and fishes), the fraction of pest species is higher—about one in four.

Inevitably, one immigrant species or another turns into a mega-invasive rivaling the frog chytrid. One such destroyer among plants is *Miconia calvescens*, an ornamental shrub native to Mexico and Central America. Two-thirds of the native forests of Tahiti have been overwhelmed by this invader. It reaches the size of trees and forms stands dense enough to crowd out all other trees and woody shrubs—and in addition all but a few small animals. A similar fate is prevented in Hawaii only by squads of volunteers who locate and weed out every individual miconia plant from uncultivated areas.

The accommodation of invasives into novel ecosystems is not cheap. By 2005, the economic costs due to invasive species in the United States alone had risen to an estimated $137 billion annually, not counting the threats they pose to native freshwater species and ecosystems.

The land birds of the Pacific Islands have been the victims of another kind of apocalyptic force. In sheer numbers of species lost, they have been the hardest hit of all vertebrate animals. The wave of extinction that began thirty-five hundred years ago with the arrival of humans in the western archipelagoes—Samoa, Tonga, Vanuatu, New Caledonia, Fiji, and the Marianas—continued nine to seven centuries ago through the colonization of the most remote islands of Hawaii, New Zealand, and Easter Island. A few of the surviving species teeter on the brink of extinction today. Two-thirds of the nonpasserine Pacific birds, however, close to one thousand species, were extinguished. Thus some 10 percent of the bird species on

Earth were wiped out during a single episode of colonization by relatively small groups of people.

Hawaii, universally acknowledged as the extinction capital of the world, had the most to lose when the Polynesian voyagers first came ashore, and with the later help of Europeans and Asian colonists, they extinguished most of its native bird species. Gone are a native eagle, a flightless ibis, a ground species the size of a turkey, and more than twenty species of drepanidid honeycreepers, the latter small pollen feeders, many with brilliantly colored plumage and long curved bills that probe deeply into tube-petaled flowers. And many more—in excess of forty-five species—vanished following the arrival of the Polynesians before A.D. 1000, and twenty-five followed the entry of the first Europeans and Asians two centuries ago. Oddly, the feathery remains of some of the most colorful extinct species are preserved in the cloaks of the old Hawaiian royalty.

The Pacific archipelagoes were a killing field for two reasons. First, because of their relatively small size and the rapid reproduction of the colonists, they were soon overpopulated. In some remote islands the predation continues on a reduced scale. In 2011, on the big island of Espiritu Santo, Vanuatu, I saw hunters armed with powerful slingshots carrying a native Pacific imperial pigeon (*Ducula pacifica*), resplendent in red-knobbed beak, white body, and black wings, on its way to a restaurant in Luganville.

The second reason for the mass extinction was that the island birds were unafraid of the two-legged colonists, having never been exposed to comparable predators during their evolution. (Snakes, mongooses, and tigers, being poor oceanic voyagers, never made it across the Pacific.) Many of the bird species had also become flightless or close to it, a common trait of land birds on small, remote islands. Hence they exemplified a basic rule of extinction biology: the first to fall are the slow, the dumb, and the tasty.

The dodo of Mauritius, an oversized, flightless descendant of a pigeon, illustrates the same principle for islands in the Indian Ocean. The first Dutch sailors to land on Mauritius, in 1598, found a bird that was just short of being served to them on a dinner platter— fat, earthbound, and fearless. The last recorded time a living dodo waddled within sight of a human being was 1662. A similar early fate was suffered by its cousin the solitaire, a native of the nearby island of Rodrigues. A third bird of a wholly different kind, the Mauritian kestrel, a small falcon, was on the brink of extinction when in 1974 the last four individuals were captured, and protected in an aviary for breeding. When descendants were populous enough to be safely released, a few individuals were returned to one of the very small surviving natural areas. The Mauritian kestrel, reduced to near oblivion by the greed of humanity, survives today but only by the tenuous grace of humanity.

In 2011, as I was conducting research on ants with a small team of fellow biologists in the mountains of the Pacific island of New Caledonia, we witnessed a dodo-like bird in the making. This strange creature, the kagu, is found only on the main island of the French province in the southwestern Pacific. Once abundant enough to be designated the official bird of New Caledonia, the kagu population is now down to fewer than a thousand individuals. The bird is a classical island inhabitant, helpless in the presence of humans, dogs, and feral cats. It is about the size of a chicken, bluish white in color, and has a prominent straight reddish bill, long pale reddish legs, and a trailing crest of white crown feathers, which are raised and spread out in dramatic display when two birds meet. Kagus live in dense upland forests. They forage on the ground, feeding mostly on insects. Although they have wings of normal size, they are able to fly only short distances.

As a typical product of island evolution, the kagu is also distressingly tame. When approached by a human, it just walks away, occasionally stopping out of sight behind a tree trunk and waiting for the intruder to leave. A student member of our group, Christian Rabeling, knew how to attract kagus. He demonstrated so with one of the birds that wandered into our path. I have no idea how Rabeling, who had never been to New Caledonia, knew what to do, but he confidently squatted and worked his hands to rustle a pile of leaves. Soon the kagu walked up to him, to examine the pile of leaves. We guessed that its reckless behavior is due to the way kagus use fellow kagus to find the insects and other invertebrates that make up their diet. Our visitor then blithely strolled away. A hunter willing to break the law could easily have grabbed it by the neck. No doubt countless New Caledonians and French colonists did so in earlier days.

Entirely different habitats that are extremely vulnerable to species extinction are streams and other bodies of water mostly small to modest in area. Just as islands are fragments of land surrounded by water, so are streams, rivers, ponds, and lakes aquatic islands surrounded by land. Freshwater species of all kinds are at high risk of extinction because on every continent except Antarctica, humans are both short of clean freshwater and in direct competition for that water with the faunas and floras living in it.

The agent causing the most immediate damage to species in fresh water are dams, great boosters of local economies but unfortunately chief demons of aquatic habitat destruction. Their malign effects include the barricades they create against migrating fish species, the stilling and deepening of water upstream, and pollution, all exacerbated by the usual intensive agricultural conversion that builds up around dams. Most at risk are salmon, sturgeons, and other fish that travel upstream to breed. One species I personally care about because of my own geographic origin is the Alabama sturgeon. It is so rare that one is caught only once every few years. Occasionally it is considered to be most likely extinct. Then another is found, amid wide publicity, and the species returns for a while to the category of "critically endangered."

For centuries the baiji, the little native dolphin of the Yangtze River, was cherished by the Chinese river people. By 2006, as the Three Gorges Dam neared completion, the baijis could no longer be found. Similar examples abound on other continents. Among the most famous is in Africa. In 2000 a hydropower plant in Tanzania's Udzumgwa Mountains cut off 90 percent of the water pouring into Kihansi Gorge, driving the tiny, golden-hued Kihansi spray toad to extinction in the wild. It survives only in a few specially designed aquariums in the United States. The plight of the little animal should all by itself serve as a wake-up call to us for mass extinctions imminent or in progress around the world.

Few Americans are aware of the damage done to the wildlife by their own dams. The greatest blow in modern history was the loss of freshwater mollusks that followed the impoundment of rivers in the Mobile River and Tennessee River basins. The toll of the Mobile River Basin in recent decades has been nineteen species of mussel (a clamlike bivalve) and thirty-seven species of aquatic snails. The loss in the Tennessee River Basin has been comparable.

In order to bring the recent mollusk losses into focus for you, I offer here the names of all nineteen of the river mussel species known to have been driven to extinction: Coosa elktoe, sugarspoon, angled riffleshell, Ohio riffleshell, Tennessee riffleshell, leafshell, yellow blossom, narrow catspaw, forkshell, southern acornshell, rough combshell, Cumberland leafshell, Apalachicola ebonyshell, lined pocketbook, Haddleton lampmussel, black clubshell, kusha pigtoe, Coosa pigtoe, stirrup shell. *Rest in peace.*

Their strangeness dramatizes how unfamiliar everywhere are the names of vanished invertebrates compared with those of extinct bird species from the same region—ivory-billed woodpecker, Carolina parakeet, passenger pigeon, Bachman's warbler.

In case this partial roster of the vanished feels unimportant to you ("What good is just another kind of river mussel?"), let me mention their practical value for human welfare. Like the oysters of bays and deltas, mussels filter and clean the water. They are a vital link in aquatic ecosystems. And in case you insist on something of immediate tangible value, they are—at least once were—a commercial source of food and mother-of-pearl.

If mussels and other invertebrates still seem of lesser relevance, let me bring fishes into the picture. From 1898 to 2006, according to Noel M. Burkhead of the American Fisheries Society, fifty-seven kinds of freshwater fish declined to extinction in North America. The causes included the damming of rivers and streams, the draining of ponds and lakes, the filling in of springheads, and pollution, all due to human activity. The rate of extinction of species and races is conservatively estimated to be 877 times above that prevailing before the origin of humanity (the latter rate is one extinction every three million years). Here, to bring them at least a whisper closer to their former existence, is a partial list of their common names: Maravillas red shiner, plateau chub, thicktail chub, phantom shiner, Clear Lake splittail, deepwater cisco, Snake River sucker, least silverside, Ash Meadows poolfish, whiteline topminnow, Potosi pupfish, La Palma pupfish, graceful priapelta, Utah Lake sculpin, Maryland darter.

Finally, there is a deeper meaning and long-term importance of extinction. When these and other species disappear at our hands, we throw away part of Earth's history. We erase twigs and eventually whole branches of life's family tree. Because each species is unique, we close the book on scientific knowledge that is important to an unknown degree but is now forever lost.

The biology of extinction is not a pleasant subject. The death of a species is especially disheartening to the scientists who study endangered and newly extinct species. Together these vanishing remnants of Earth's biodiversity test the reach and quality of human morality. Species brought low by our hand now deserve our constant attention and care. Religious believers and nonbelievers alike would do well to sacralize God's elegant command given in the Judeo-Christian account of Genesis: *Let the waters teem with countless living creatures, and let birds fly above the earth across the vault of heaven.*

Australian bustard in courtship display.
Proceedings of the Zoological Society of London, *1868.*

6

ARE WE AS GODS?

S ome believe that humanity should accept the ecological chaos we have created as just damaged collateral to a brilliant destiny. "We are as gods," the futurist Stewart Brand has written, "and have to get good at it." Earth is our planet, this vision continues in zigzag logic, and our ultimate role is to take control of all of it. A few kerfuffles such as economic crashes, climate change, and religious wars notwithstanding, we are getting better in every way all the time. We travel ever faster around the globe, reach higher and probe deeper, and look farther across the universe. We are collectively learning at an exponential pace everything the Big God permits us little gods to learn, and are putting all that knowledge within reach of everyone by a few keystrokes. We are the pioneers of a whole new kind of existence. *Homo sapiens*, the amazing primate species on two legs, with free hands and a cerebrum-packed globular skull, is on its way!

In the dreams of science-prone intellectuals and Hollywood scriptwriters, there is no limit to humanity's eventual grasp.

Astrophysicists conceive of travel at one-tenth the speed of light among the two hundred billion stars of the Milk Way, traversing it if desired in a few tens of thousands of years. There is even time enough ahead for a species like our own to occupy the galaxy. The formula is arithmetically simple, as follows. Take centuries to colonize the nearest star system that possesses a habitable planet. Take a few more centuries or millennia to build a civilization there, then launch multiple space vehicles to other star systems. Continue the process until all the habitable planets of the galaxy are occupied. The time may seem impossibly long but it is less than that between the evolutionary origin of humans and the present (which also is only an eyeblink in the full span of life on Earth).

Meanwhile, in imagination, we may attain the status of what the astronomer Nikolai Kardashev called Type I civilization, a society in control of all the available energy on Earth. Thus we conceivably could press on to Type II civilization, in control of available power in the Solar System, and even Type III civilization, taking control of all energy in the galaxy.

May I now humbly ask, just where do we think we are going— really? I believe the great majority of people on Earth would agree with the following goals: an indefinitely long and healthy life for all, abundant sustainable resources, personal freedom, adventure both virtual and real on demand, status, dignity, membership in one or more respectable groups, obedience to wise rulers and laws, and lots of sex with or without reproduction.

There is a problem, however. These are also the goals of your family dog.

Let's talk about ourselves. We are indeed somehow soaring to greatness, if not godlike at least for our own emotional gratification. Our individual organismic selves, our tribe, our species, are the culmination of Earth's achievement. Of course we think this way. So would members of any other species capable of self-reflection at the human level. If it could think, each and every fruit fly would yearn for greatness. We are so brainy compared with the rest of life that we actually do think of ourselves as demigods, somewhere halfway between the animals below us and angels above, and moving ever upward. It is easy to suppose that the genius of our species is on some kind of automatic pilot, guiding us to an undefined empyrean that will exist with perfect order and provide personal happiness. If we ourselves are ignorant, our descendants will find the empyrean as humanity's destiny when someday, somehow, they arrive there.

So we stumble forward in hopeful chaos, trusting that the light on the horizon is the dawn and not the twilight. Ignorance of the future based on lack of self-understanding is, however, a dangerous condition. The French writer Jean Bruller had it right when on the brink of World War II he wrote that "all of mankind's troubles are due to the fact that we do not know what we are and cannot agree on what to become."

We are still too greedy, shortsighted, and divided into warring tribes to make wise, long-term decisions. Much of the time we behave like a troop of apes quarreling over a fruit tree. As one consequence, we are changing the atmosphere and climate away from conditions best for our bodies and minds, making things a lot more difficult for our descendants.

And while at it, we are unnecessarily destroying a large part of the rest of life. Imagine! Hundreds of millions of years in the making, and we're extinguishing Earth's biodiversity as though the species of the natural world are no better than weeds and kitchen vermin. Have we no shame?

In order to settle down before we wreck the planet, we should at the very least learn to think about where our species really came from and what we are today. There is plenty of evidence to show that transcendent goals—above self and tribe—do arise in the human brain. They are fundamentally biological in origin. To understand the meaning of life, to know that we know and how and why we know, is the premier driving force of all of science and the humanities. There is greatness in understanding the basic elements of human evolution and wisely acting upon the way they are linked. The form it is taking can be expressed succinctly as follows: the biosphere gave rise to the human mind, the evolved mind gave rise to culture, and culture will find the way to save the biosphere.

For those who must believe in gods (and who can say with provable certainty they are wrong?), we may hope it will not be like the bloodthirsty warrior god described in Joshua 10. His name was Yahweh, and He halted the celestial sphere in order to aid the Amorite genocide, guaranteeing Israelite victory. On behalf of His people He commanded:

Stand still, O Sun, in Gibeon;
stand, Moon, in the Vale of Aijalon.*

May the better instruction be instead the one by Paul, in his first letter to the Corinthians, advising them to look inward for wisdom from the Lord of glory, and hence seek:

Things beyond our seeing, things beyond our hearing,
things beyond our imagining, all prepared by God for those
who love him.*

There is an unbreakable chain in self-understanding that thinking people largely neglect. One of its lessons is that we are not as
gods. We're not yet sentient or intelligent enough to be much of
anything. And we're not going to have a secure future if we continue to play the kind of false god who whimsically destroys Earth's
living environment, and are pleased with what we have wrought.

* New English Bible.

Thylacine ("Tasmanian tiger") of Australia, extinct 1936. Proceedings of the
Zoological Society of London, *1848–1860.*

7

WHY EXTINCTION IS ACCELERATING

Very few people wish to see species disappear, except those occasional pests that attack our bodies and sources of food. The biosphere would not mourn the loss of mosquitoes of the African *Anopheles gambiae* group of species, specialists on human blood, expert hiders in native dwellings, and the principal carriers of malaria. Nor, I suspect, would even dedicated conservationists mourn the complete elimination of the African guinea worm, my candidate for the most ghastly of human pathogens. Growing up to a meter in length, it stretches through the body and opens sores on the feet or legs to eject its larvae. We could moreover endure extinction of the protozoan parasites that cause the disfiguring and deadly disease of leishmaniasis. Other than still-unknown pathogens among bacteria, microscopic fungi, and viruses, the number of species worthy of extinction, or at least of harmless storage in liquid nitrogen, is probably (my guess) fewer than a thousand. I have suffered several times from arboviruses (arthropod-borne) picked up in tropical forests, lying abed with fever, and would happily say goodbye to them.

The millions of other species are beneficial to human welfare, whether directly or indirectly. There are unfortunately almost countless ways that humans are hastening their extinction, whatever might be their present or future beneficent roles. The human impact is largely due to the excess of the many quotidian activities we perform just to get on with our personal lives. Those activities have made us the most destructive species in the history of life.

How fast are we driving species to extinction? For years paleontologists and biodiversity experts have believed that before the coming of humanity about two hundred thousand years ago, the rate of origin of new species per extinction of existing species was roughly one species per million species per year. As a consequence of human activity, it is believed that the current rate of extinction overall is between one hundred and one thousand times higher than it was originally, and all due to human activity.

In 2015 an international team of researchers finished a careful analysis of the prehuman rates and came up with a diversification rate ten times lower in genera (groups of closely related species). The data, when translated to species extinctions, suggests species extinction rates at the present time are closer to one thousand times higher than that before the spread of humanity. The estimate is further consistent with an independent study that detected a similar downward shift in the rate of species formation in prehumans, as well as in their closest relatives among the great apes.

Every expansion of human activity reduces the population size of more and more species, raising their vulnerability and the rate of extinction accordingly. A 2008 mathematical model by a team of botanists predicted that between 37 and 50 percent of rare tree species in the Brazilian Amazon rain forest, "rare"

defined as having populations of fewer than ten thousand individuals, will suffer early extinction, caused by contemporary
road building, logging, mining, and conversion of land to agriculture. The lower figure, 37 percent, applies to areas developed
in part but protected by careful management.

It is difficult to make comparisons of origin and extinction rates
across different kinds of plants and animals in different parts of
the world. But all of the available evidence points to the same two
conclusions. First, the Sixth Extinction is under way; and second,
human activity is its driving force.

This grim assessment leads to a second very important question:
How well is conservation working? How much have the efforts of
global conservation movements achieved in slowing and halting the
devastation of Earth's biodiversity? Having served on the boards
of Conservation International, The Nature Conservancy, and the
World Wildlife Fund—U.S., and as advisor to many local conservation organizations, I can testify to the zeal and inspiration, backed
by private and public funding, and to the years of sweat and blood
in the field, that have gone into conservation efforts around the
world during the past half century. How much has this heroic effort
accomplished?

In 2010 a survey conducted by close to two hundred experts on
vertebrate land animals (mammals, birds, reptiles, amphibians) analyzed the status of all the 25,780 known species. One-fifth were con-
firmed as threatened with extinction, and of these a fifth had been
stabilized as a result of conservation efforts. An independent study
in 2006 had already concluded that extinction of bird species in particular had been cut by about 50 percent as a result of conservation
efforts during the past century. Thirty-one bird species worldwide

still live because of efforts on their behalf. In short, global conserva-tion thus far, when averaged out for land-dwelling vertebrates, has lowered extinction rates of species by approximately 20 percent.

Next, what is the impact of governmental regulation, in particular the U.S. Endangered Species Act of 1973? A review made in 2005 found that a quarter of the 1,370 American plant and animal species classi-fied earlier as threatened achieved new population growth, while 40 percent declined, with 13 of the listed species improved enough to be taken off the endangered list. The most important statistic is that while 22 species had slipped into extinction, 227 had been saved that would likely have otherwise disappeared. Among the more familiar protected species that have climbed back to health are the yellow-shouldered blackbird, green sea turtle, and bighorn sheep.

These successes have shown that conservation works, but at the level of effort being applied at the present, it falls far short of what is needed to save the natural world. The conservation movement has slowed the species extinction rate but failed to bring it anywhere close to the prehuman level. At the same time the birth rate of species is dropping rapidly. Like an accident patient in emergency care continuing to hemorrhage and with no new supply of blood available, stabilization is out of reach and further decline and death are inevitable. We might be inclined to say to the surgeons and conservationists alike, "Congratulations. You have extended a life, but not by much."

Of course, not all wild species are threatened by the assault on biodiversity. A few are compatible with a humanized environment. What fraction of the present survivors will last to the end of this century? If present conditions persist, perhaps half. More likely fewer than one-fourth.

That is my guess. But the fact is that due to habitat loss alone, the rate of extinction is rising in most parts of the world. The preeminent sites of biodiversity loss are the tropical forests and coral reefs. The most vulnerable habitats of all, with the highest extinction rate per unit area, are rivers, streams, and lakes in both tropical and temperate regions.

An established principle in conservation biology for all habitats is that a reduction in area results in a fraction of the species disappearing in time by roughly the fourth root of the area. If 90 percent of a forest is cut, for example, about half of the species will soon disappear that would have otherwise persisted. At the beginning most of the species may survive for a while, but roughly half will have populations too small to persist for more than a few generations.

Barro Colorado Island in Panama has proved a valuable natural laboratory to study the effect of area on extinction. Covered by rain forest, it was created by the formation of Gatun Lake in 1913 during the construction of the Panama Canal. An ornithologist, John Terborgh, predicted that after fifty years the island would lose seventeen bird species. The actual number was thirteen, representing 12 percent of the one hundred eight breeding species originally present. On the other side of the world, a 0.9-square-kilometer patch of rain forest, the Bogor Botanical Gardens of Indonesia, was also isolated, not by water but by the clearing of all the forest around it. In the first fifty years it lost twenty of its sixty-two breeding bird species, approximately the number expected.

Conservation scientists often use the acronym HIPPO for a quick recall of the most ruinous of our activities, in order of importance:

Habitat destruction. This includes that caused by climate change.

Invasive species. This includes plants and animals that crowd out native species and attack crops and native vegetation, as well as microbes causing disease in humans and other species.

Pollution. The effluents from human activity are killers of life, especially in rivers and other freshwater ecosystems, the most vulnerable of Earth's habitats.

Population growth. Although it is still widely unpopular to say so, we must really slow down. Reproduction is obviously necessary, but it is a bad idea, as Pope Francis I has pointed out, to continue multiplying like rabbits. Demographic projections suggest that the human population will rise to about eleven billion or slightly more before the end of the century, thereafter peak, and begin to subside. Unfortunately for the sustainability of the biosphere, per-capita consumption is also destined to rise, and perhaps even more steeply than human numbers. Unless the right technology is brought to bear that greatly improves efficiency and productivity per unit area, there will be a continued increase in humanity's ecological footprint, defined as the area of Earth's surface each person on average needs. The footprint is not just local area, but space scattered across land and sea, in pieces for habitation, food, transportation, governance, and all other services down to and including recreation.

Overhunting. Fishing and hunting can be pressed until the target species is driven to extinction or on the way, making the last surviving populations subject to final erasure by disease, competition, changes in weather, and other stresses survived by larger and more far-ranging populations of the same species.

Single causes in the decline and extinction of species can be identified readily in a few cases. One example is the dietary preferences of the brown tree snake, which is especially skilled at preying on bird nestlings. Another has been identified in the decline of the monarch butterfly in the Midwest, which famously overwinters in masses of millions on pine trees in the Mexican state of Michoacán. By 2014, there was an 81 percent decline of the butterflies in the United States Midwest population, attributed to a 58 percent decline in milkweeds, the exclusive food plant of the monarch caterpillars. The milkweeds have fallen off in turn because of the increased use of glyphosate weed-killer in corn (maize) and soybean fields. The crops have flourished after being genetically modified to resist the weed-killer, while the wild milkweed plants have not been so protected. With their food supply unintentionally reduced, the migrating monarch butterflies of both the United States and Mexico have declined steeply.

In most extinctions, however, the causes are multiple, linked to one another in some way, all ultimately the result of human activity. In one well-analyzed example of multiple causes, the Allegheny woodrat has vanished or grown endangered through a third of its range. It is considered to have suffered from the extinction of the American chestnut and hence the loss of its seeds, on which the species partly depended. Also important have been the logging and fragmentation of the forests where the woodrats live, plus further reduction of this habitat by the voracity of the invasive European gypsy moth. The coup de gras is roundworm infection from raccoons, which are animals better suited than the woodrats to live around humans.

Those unimpressed by the fall of a rodent may turn a more caring eye to the songbirds that migrate each year between their wintering ground in the New World tropics and breeding range in the eastern United States. From data compiled by the federally sponsored North American Breeding Bird Survey along with the Audubon Christmas Bird Count, it is clear that populations of more than two dozen species are in a steep descent. Those affected include the wood thrush, Kentucky warbler, eastern kingbird, and bobolink. One, Bachman's warbler, whose wintering ground was in Cuba, has evidently gone extinct. I have a place in my heart for this little bird. During field trips to the floodplain forests of the U.S. Gulf Coast, which have brought me close to canebrakes where the warbler once nested, I've often looked and listened for a Bachman's as best I could (admittedly not very skillfully), but to no avail.

It sometimes seems as though the remainder of American native plants and animals are under deliberate assault by everything humanity can throw at them. Leading the list in our deadly arsenal are the destruction of both wintering and breeding habitats, heavy use of pesticides, shortage of natural insect and plant food, and artificial light pollution causing errors in migratory navigation. Climate change and acidification pose newly recognized yet game-changing risks—shifts in all aspects of the rhythms of the environment away from those necessary for wildlife survival and reproduction.

There exist several facts about global biodiversity to keep in mind while trying to save it. The first is that the human-caused agents of extinction are synergistic. As any one of the agents intensifies, it causes others to intensify also, and the sum of the changes is an acceleration of extinction. Clearing a forest for agriculture reduces habitat, diminishes carbon capture, and introduces pollutants that are carried downstream to degrade otherwise pure aquatic habitats en route. With the disappearance of any native predator or herbivore species, the remainder of the ecosystem is altered, sometimes catastrophically. The same is true of the addition of an invasive species.

Another overarching principle of biodiversity is the greater richness of tropical environments over temperate environments, in both number of species and vulnerability. While the variety of aphids, lichens, and conifers increases poleward, a vastly larger number of other kinds of organisms increase in the opposite direction. For example, you can expect to find about fifty species of ants in a square kilometer of New England temperate forest (if you care to look) but up to ten times that number in a comparable area of rain forest in Ecuador or Borneo.

A third principle of biodiversity worth noting is in the relation between its richness and the geographical range of the species comprising it. A large fraction of the species of plants and animals of temperate North America are distributed across most of the continent, but very few species range the same way across tropical South America.

When these last two principles are linked, both entailing the numbers of resident species, we find as expected that on average tropical species are more vulnerable than temperate species. They occupy smaller ranges and thereby manage to sustain smaller populations. Furthermore, existing as they do among a greater array of competing species, they tend to be more specialized in where they live, in what they eat, and by the predators that hunt and eat them.

A general rule to follow in conservation practice is therefore that, while clear-cutting a square kilometer of old-growth coniferous forest in Canada, Finland, or Siberia will do a lot of environmental damage, cutting the same area of old-growth rain forest in Brazil or Indonesia will do far more damage.

Finally, there is the immense disparity between the 62,839 known species of vertebrates (mammals, birds, reptiles, amphibians, fishes, the total number in 2010) and 1.3 million known invertebrate species (also 2010). Almost all the information on quantitative trends in biodiversity are based on the vertebrates, the big animals with which we have intimate familiarity. There are well-studied groups within the invertebrates; notably including mollusks and butterflies, but even these organisms are less well known than the mammals, birds, and reptiles. The great majority of invertebrate species, notably the hyperdiverse insects and marine organisms, remain to be discovered and made known to science. Nevertheless, of those groups well enough studied to make an estimate of their conservation status down to species, such as freshwater crabs, crayfish, dragonflies, and corals, the percentages of vulnerable and endangered species are comparable to those of vertebrates.

In thinking about life and death in the biosphere, it is important to avoid two misconceptions. The first is that a rare declining species is probably senescent. You might think that its time has come, so let it go. On the contrary, its young are just as vital as the young of the most aggressive expanding species with which they compete. If its population is shrinking in size from vulnerable to endangered to critically endangered (the scale of descent used in the Red List of the International Union for Conservation of Nature),* the reason is neither age nor destiny of the species. Instead, the reason is the predicament in which the Darwinian process of natural selection has placed it. The environment is changing, and the genes assembled by earlier natural selection are by happenstance not able to adapt quickly enough. The species is a victim of bad luck, rather like a

---

* The Red List scale for individual species as of 2001 is the following in descending order: Least Concern (LC), Near Threatened (NT), Endangered (EN), Critically Endangered (CR), Extinct in the Wild (EW), Extinct (E).

farmer who invests in land at the start of a ten-year drought. Put some young individuals into an environment where the genes are better adapted, and the species will flourish.

Humanity, keep in mind, is the principal architect of such maladaptive environments. Conservation biology is the scientific discipline by which better environments are identified and protected or restored for species imperiled within them.

Biologists recognize that across the 3.8-billion-year history of life over 99 percent of all species that lived are extinct. This being the case, what, we are often asked, is so bad about extinction? The answer, of course, is that many of the species over the eons didn't die at all, they turned into two or more daughter species. Species are like amoebae; they multiply by splitting, not by making embryos. The most successful are the progenitors of the most species through time, just as the most successful humans are those whose lineages expand the most and persist the longest. The birth and death rates of humans are close to a global balance, with birth having the edge for the last sixty-five thousand years or so. Most important, we, like all other species, are the product of a highly successful and potentially important line that goes back all the way to the birth of humanity and beyond that for billions of years, to the time when life began. The same is true of the creatures still around us. They are champions, each and all. Thus far.

Starfish and tubeworms. Alfred Edmund Brehm, 1883–1884.

8

THE IMPACT OF CLIMATE CHANGE:
LAND, SEA, AND AIR

Having risen above all the biosphere, set to alter everything everywhere, the wrathful demon of climate change is our child that we left unrestrained for too long. By using the atmosphere as the carbon dump of the Industrial Revolution, and pressing on without caution, humanity has raised the concentration of greenhouse gases, primarily carbon dioxide and methane, to a dangerous level.

Most experts agree on the following dire prediction. The rise in the annual mean surface temperature caused by the pollution should not be allowed to exceed 2°C, or 3.6°F, above that prior to the birth of the Industrial Revolution—roughly, the mid-eighteenth century. The rise has already reached nearly half of the "2C" threshold. When global atmospheric warming pushes past the 2C-increase level, Earth's weather will be destabilized. Heat records now considered historic will become routine. Severe storms and weather anomalies will be the new normal. The melting of the Greenland and Antarctic ice shields now under way will accelerate, bringing landmasses a new climate and a new geography. The sea level, as measured by both satellite and tidal gauge data, is already rising three millimeters a year. Pushed up both by the addition of melt-water and by expansion of the ocean volume due to heating of the whole of marine water itself, the sea level will eventually exceed nine meters.

Can such catastrophic change really come to pass? It has already begun. The average annual surface temperature of the planet has increased steadily since 1980, with no sign of moderating.

The governments of the world have been stirred to action, but the response is tepid and far from adequate. Only the tiny Pacific Island countries of Kiribati and Tuvalu, with the Pacific Ocean threatening to close over them, have found a solution. They are get-ting ready to move their populations to New Zealand.

Of course, you don't see the moving average from one day to the next. Political leaders in Washington haven't begun yet to travel to their offices in gondolas. But on November 12, 2014, President Obama did sign a historic agreement with President Xi Jinping of China that requires the United States to reduce its carbon emis-sions to 28 percent below the 2005 level by 2025, while China will peak and start downward to the same level by 2030. In December 2014, delegates of 196 sovereign nations, virtually from the whole world, met in Lima, Peru. They agreed to return home and com-pose a plan within six months for their respective countries to cut their own emissions of greenhouse gases from coal, gas, and oil. The collective plan was the basis of a global agreement to be drafted during December in 2015. The accord reached, however, will not have to be implemented until 2020.

The International Energy Agency insists that somehow humanity must plan to leave most of the world's proven oil and gas reserves in the ground to blunt otherwise ruinous climate change, adding that "no more than one-third of proven reserves of fossil fuels can be consumed prior to 2050."

A dilemma facing each one of the delegates lies in the division of loyalties. Called by the late ecologist Garrett Hardin the tragedy of the commons, it originates when individuals, or organizations or nations, share a limited resource. They will tend to deplete the resource—in this case global clean air and water—because each will obtain as large a share as allowed by the rules or an even larger share by outright cheating.

A textbook example of the tragedy of the commons is the depletion of living resources in the open sea. Within territorial waters the harvesting of fish and other seafood is weakly regulated in many parts of the world, if at all. Blue water, belonging to no one, is subject to no regulations whatsoever, save that established by international negotiation. For generations all marine waters, variously protected to some degree or not at all, have suffered overharvesting of edible species. The downward spiral has been hastened by habitat destruction, spread of invasive species, climate warming, acidification, pollution with toxins, and eutrophication from excess nutrient runoff.

The assault has been brutal and unrelenting. The numbers of larger food and sports species, such as tuna, swordfish, sharks, and larger groundfish (the latter comprising cod, halibut, flounder, red snapper, and skates), have fallen 90 percent since 1950. Cod, so abundant in the time of the Pilgrims that it was said a meal could be caught with an unbaited hook, have declined by over 99 percent.

Fortunately, the complete extinction of ocean species is much rarer than that of larger animals on the land. The reason is that almost all of the larger marine species, including pelagic fishes, have large personal home ranges or migratory routes. In a lifetime they travel much greater distances than larger land animals, thus seeding pauperized populations—and hence saving their species from extinction. Where Asian tigers, for example, have been extirpated from about 93 percent of their original geographical ranges, tiger sharks still roam through almost all of theirs.

Unfortunately, the opposite is true of coral reefs. Often called the "rain forests of the sea" for their seemingly endless biodiversity, the reefs are an outstanding exception to the resilience of marine ecosystems. Corals are symbiotic organisms. Each coral consists of a calcareous, plantlike animal, which is the part you see. It contains large numbers of single-celled microorganisms called zooxanthellae, which you don't see except through their vibrant colors. The coral skeleton creates the architecture of a reef in the same way trees and shrubs create the architecture of a forest. The zooxanthellae are photosynthesizers that provide the energy and substance with which the calcareous structure is built.

With as little as 1°C warming of the water, or a small increase in its acidity, both of which are caused by human action, the zooxanthellae emigrate from their host calcareous organism, taking their colors and photosynthesis with them, the potentially suicidal process called coral bleaching.

For these magnificent coral assemblages the consequences of changes caused by warming are already catastrophic. Nineteen percent of the world's coral reefs are dead. Thirty-eight percent of the 44,838 coral species known in the world are vulnerable or endangered, compared to 14 percent of the birds, 22 percent of the mammals, and 31 percent of the amphibians (frogs, salamanders, and caecilians). Recent analyses suggest that a quarter of the world's coral species will vanish by 2050.

Colonial fruit bats ("flying foxes"), Old World tropics.
Alfred Edmund Brehm, 1883–1884.

9

THE MOST DANGEROUS WORLDVIEW

Not everyone claiming to be a conservationist agrees that biodiversity should be protected intact. A small but growing minority believe that humanity has already changed the living world beyond repair. We must now, they say, adapt to life on a damaged planet. A few of the revisionists urge the adoption of an extreme Anthropocene worldview, in which humans completely dominate Earth and surviving wild species and ecosystems are judged and conserved for their usefulness to our species.

In this vision of life on Earth, wildernesses no longer exist; all parts of the world, even the most remote, have been adulterated to some degree. Living nature, as it evolved before the coming of man, is dead or dying. Perhaps, extreme advocates believe, this outcome was foreordained by the imperatives of history. If so, the destiny of the planet is to be completely overtaken and ruled by humanity—pole to pole, of, by, and for us, the only species at the end of the day that really matters.

There is a whisper of truth in this opinion. Humanity has delivered a blow to the planet not even remotely approached by that of any other single species. The full scale of the assault, in common parlance of the Anthropocene called "growth and development," began at the start of the Industrial Revolution. It was foreordained by the extermination of most of the mammal species in the world more than ten kilograms in weight, collectively called the megafauna, a process begun by Paleolithic hunter-gatherers and thereafter increased in stages enabled by technological innovation.

The decline of biodiversity has been more like the gradual dimming of light than the flick of a switch. As the human population multiplied and spread around the world, it almost always strained local resources to their local limits. Doubling in numbers, then doubling again, and yet again, people fell upon the planet like a hostile race of aliens.

The process was pure Darwinian, obedient to the gods of unlimited growth and reproduction. While the creative arts yielded new forms of beauty by human standards, the overall process has not been pretty by anybody else's standards—except bacteria, fungi, and vultures. As described in 1877 by the Victorian poet Gerard Manley Hopkins:

> Generations have trod, have trod, have trod;
> And all is seared with trade; bleared, smeared with toil;
> And wears man's smudge and shares man's smell; the soil
> Is bare now, nor can foot feel, being shod.

The extirpation of biodiversity proceeded in a measure equal to that of the spread of mankind. Tens of thousands of species fell to the ax and the pot. As we've already seen, at least a thousand species of birds, 10 percent of the world total, vanished when the Polynesian colonists swept across the Pacific on double canoe and outriggers island to island from Tonga to the most distant archipelagoes of Hawaii, Pitcairn, and New Zealand. The early European explorers of North America found that the megafauna, possibly once the richest in the world, had already been wiped out by the arrows and traps of the Paleoindians. Gone were mammoths, mastodons, saber-toothed big cats, oversized dire wolves, huge soaring birds and gigantic beavers and earthbound sloths.

Yet in the most pauperized regions, most of the plants and smaller animals, including the always hyperdiverse insects and other arthropods, remained largely intact. I'm confident that if I could go back in time fifteen thousand years with net and shovel, I could find and identify most of the kinds of butterflies and ants. Yet the megafauna would be an entirely new world. The conservation movement, born in the United States during the nineteenth and early twentieth centuries, came late, but mercifully not too late to save what is left of our fauna and flora. Starting with the Yellowstone National Park in 1872, first of its kind in the world, and inspired by the writings of Henry David Thoreau, John Muir, and other naturalists and activists, the movement has culminated in impressive networks of federal, state, and local parks. These are augmented by private reserves set aside by nongovernmental organizations such as most notably The Nature Conservancy. Nature is wild, nature is ancient, nature is pure, this essentially American credo declared, and nature should not be managed except to blunt the corroding effects of human interference. America's national parks are "the best idea we ever had," the author Wallace Stegner wrote in 1983.

The concept of conservation for its own sake has spread around the world, to the extent that by the early twenty-first century a great majority of the world's 196 sovereign national states had natural national parks or government-protected reserves of some kind. The concept has therefore been successful—but only partly so in numbers and quality. Critically endangered wetlands—harboring species tenfold more in numbers than those in the American and European reserves—barely hang on across large swaths of tropical America, Indonesia, the Philippines, Madagascar, and equatorial Africa. The species extinction rate in all such habitats around the world, estimated from data on vertebrates (mammals, birds, reptiles, frogs and other amphibians, fishes), has reached about a thousand times the prehuman baseline, and is accelerating.

The shortfall of the conservation movement has been the focus of the new Anthropocene ideology. Its proponents claim in essence that traditional efforts to save Earth's biodiversity have failed. Pristine nature no longer exists, and true wildernesses survive only as a figment of the imagination. Those who see the world through the lens of Anthropocene enthusiasts work off an entirely different worldview from that of traditional conservationists. Extremists among them believe that what is left of nature should be treated as a commodity to justify saving it. The surviving biodiversity is better judged by its service to humanity. Let history run its seemingly predetermined course. Above all, recognize that Earth's destiny is to be humanized. For those sharing most or all of this vision, the Anthropocene is intrinsically a good thing. What remains of nature is not bad, of course, but the bottom line remains that even wildlife must earn its livelihood like everyone else.

The ideology, which some of its proponents call the "new conservation," has led to a variety of practical recommendations. First and foremost, nature parks and other reserves should be managed in a way that helps them meet the needs of people. And not all people, but implicitly those of us alive at the present time and into the near future, with our contemporary aesthetics and personal values made decisive—hence forever. Leaders who follow Anthropocene guidelines will take nature past the point of no return, whether the countless generations to come like it or not. Surviving wild species of plants and animals will live in a new amity with humans. Where in the past people entered natural ecosystems as visitors, in the Anthropocene Era species composing adulterated fragments of the ecosystems are expected to live among us.

Leading Anthropocene enthusiasts seem unconcerned with what the consequences will be if their beliefs are played out. They are as free of fear as they are of facts. Of them the social observer and conservationist Eileen Crist has written:

> Economic growth and consumer cultures will remain the leading social models (many Anthropocene promoters see this as desirable, while a few are ambivalent); we now live on a domesticated planet, with wilderness gone for good; we might put ecological gloom-and-doom to rest and embrace a more positive attitude about our prospects on a humanized planet; technology, including risky, centralized, and industrial-scale systems, should be embraced as our destiny and even our salvation.

Erle Ellis, an environmental scientist at the University of Maryland, has posted a fierce credo meant to help environmentalists prepare for the new order: "Stop trying to save the planet. Nature is gone. You are living on a used planet. If this bothers you, get over it. We now live in the Anthropocene—a geological epoch in which Earth's atmosphere, lithosphere, and biosphere are shaped primarily by human forces."

What passion drives the human juggernaut? It is manifested in the ordinary experience of our daily lives and in the unquestioned idioms of our language. Crist continues her analysis:

> Takeover (or assimilation) has proceeded by biotic cleansing and impoverishment: using up and poisoning the soil; making things killable; putting the fear of God into the animals such that they cower or flee in our presence; renaming fish "fisheries," animals "livestock," trees "timber," rivers "freshwater," mountaintops "overburden," and seacoasts "beachfront," so as to legitimize conversion, extinction, and commodification "ventures."

Anthropocene enthusiasts are, of course, not entirely lacking in ideas on how to preserve biodiversity in the new order. Chris D. Thomas, a conservation biologist at the University of York in the United Kingdom, has managed to work around masses of published contrary evidence to posit that the ongoing extinction of local native species will be balanced by alien species now being spread by people around the world. These, this expert assures us, will help to fill gaps in ecosystems that are either naturally low in biodiversity or else pauperized by human activity. Hybridization between the aliens and the surviving native species will further bolster the number of varieties and species. And we must keep in mind, he reminds us, that mass extinctions during past geological ages were followed by surges of new species. Of course, the process took millions of years. It does not seem to matter to Thomas that future generations will be peeved to learn that the evolutionary restoration of biodiversity will require an amount of time five million years or longer, several times that of the span needed to evolve the modern human species. Nor is it a major impediment that a significant fraction of alien species turn into invasives, costing the world many billions of dollars annually.

If the preservation of Earth's living heritage depends foremost on just leaving the heart of biological conservation in place safely, who could think differently? One prominent person who thinks differently has been Peter M. Kareiva, a leading light of the "new conservation" philosophy. He acquired an influential pulpit as chief scientific officer of The Nature Conservancy in 2014. In public speaking and writing for the academic and popular press, he has been the leader of those who attack the existence of wilderness. In his opinion there are no pristine areas left on Earth. The regions they long ago occupied should therefore be opened to people for more sensible management and profit. Instead of wildernesses, Kareiva prefers "working landscapes," presumably as opposed to "lazy and idle" landscapes, thereby making them more acceptable to economists and business leaders.

But this assault on wildernesses is based on an etymological error. Nowhere in the U.S. Wilderness Act do words like "pristine" appear. Surely Kareiva and others of like mind are also aware that the word "wilderness" refers to undomesticated places not yet yoked to the human will. In the parlance of conservation science, "wilderness" means a large area within which natural processes unfold in the absence of deliberate human intervention, where life remains "self-willed." Wildernesses have often contained sparse populations of people, especially those indigenous for centuries or millennia, without losing their essential character. And areas of wilderness, as I will document shortly, are real entities. They cannot be defined out of existence.

Other Anthropocene optimists have a different kind of hope: many extinct species, they believe, can be brought back to life, providing we can obtain enough preserved tissue to map their genetic codes and clone entire organisms. Examples favored for de-extinction, as this process is called, include passenger pigeons, mammoths, and Australia's wolflike thylacine. Presumably, the ecosystems that they require to survive will be intact, or else can be re-created, for each species, with the original niches somehow made available.

Subrat Kumar, a biotechnology professor in Bhubaneswar, India, writing in *Nature*, not only believes in de-extinction but urges a major new program to prepare for resurrection on a Noachian scale. To those who worry that the once-extinct species might spread so successfully as to wipe out other species as they rage zombie-like through the wildlands, Kumar offers a reassuring addendum: "Any species we bring back could be engineered to be eliminated easily should it pose a problem."

Meanwhile, over in the realm of popular literature, Emma Marris, journalist and author, offers a cheerful picture of semiwild species kept for human benefit in places scattered about in the gardens of a new smart planet. In her view, we should abandon forthwith the idea of untrammeled wildernesses, a "cult" born and peddled in America that has "lurked behind conservation organization mission statements" and unfortunately has managed to "saturate nature writing and nature documentaries." Such fallacious thinking must be tempered, Marris warns. Our true role as rulers of the planet is to turn its biodiversity into a "global, half-wild rambunctious garden tended by us."

It has been my impression that those most uncaring and prone to be dismissive of the wildlands and the magnificent biodiversity these lands still shelter are quite often the same people who have had the least personal experience with either. I think it relevant to quote the great explorer-naturalist Alexander von Humboldt on this subject, as true in his time as it is in ours: "The most dangerous worldview is the worldview of those who have not viewed the world."

PART II

The Real Living World

A large part of biodiversity still exists in both species and ecosystems, but the time that remains to save it is running out fast. It can be largely gone by the end of the century. What follows is an image of its immense surviving breadth.

Marine mollusks. Proceedings of the Zoological Society of London, *1848–1860.*

10

CONSERVATION SCIENCE

L ike most mistaken philosophies, the Anthropocene world-view is largely a product of well-intentioned ignorance. Its call for a new, human-centered approach to conservation—more precisely anti-conservation—has multiple sources. First is a false image of the history of conservation organizations. Second is an inadequate grasp of the biodiversity database. A third, less obvious source is the mistaken emphasis on ecosystems as the key level of biological organization, to the near exclusion of species and genes.

To suggest—as hard-core believers in anthropocentrism do in promoting the "new conservation"—that the agendas of the traditional conservation organizations give scant attention to the welfare of people is simply wrong. From personal experience during thirty years of service on the governing and advisory boards of several leading global organizations, I know very well that the opposite is true. I was present, for example, when, during the 1980s, the World Wildlife Fund–U.S. radically broadened its guidelines. We began by debating the intent and purpose of the organization. Which groups of animals and plants to protect, which parts of the world, and how, and, finally, why? Was it enough to stay with a few charismatic animals and plants, trusting they would function as "umbrella species" that shelter the rest of the life around them? And how might saving the big and beautiful of the natural world also serve humanity? Surely it was wrong—and ultimately futile—to fence off nature reserves from people already living in them or close by.

Our solutions were adopted in two steps. First, we expanded our attention from the panda-and-tiger class of star species to include entire ecosystems, even those without species familiar to the public. Next, we set out as a policy to assist the economy and health care of people living in and around the nature reserves.

Other conservation organizations similarly have altered their activities to bring humanity to center stage. Conservation International, for example, has put a focus on assisting government leaders in developing countries, suggesting ways to protect biodiversity as part of improvement of the economy and quality of life of its rural people. The Nature Conservancy has always been people-minded by assuming stewardship of biologically rich land and opening it to the public, including ecology and biodiversity researchers. Rare, a successful small conservation organization, has focused on star species and natural ecosystems as a vital part of the culture of local people.

Leaders in biodiversity research and conservation have long understood that the surviving wildlands of the world are not art museums. They are not gardens to be arranged and tended for our delectation. They are not recreation centers or harborers of natural resources or sanatoriums or undeveloped sites of business opportunities—of any kind. The wildlands and the bulk of Earth's biodiversity protected within them are another world from the one humanity is throwing together pell-mell. What do we receive from them? The stabilization of the global environment they provide and their very existence are the gifts they give to us. We are their stewards, not their owners.

A prediction can't yet be made of the damage from procedures suggested by Anthropocene ideologists, in particular their half-wild gardens, alien and new hybrid species, and business-friendly landscapes. The meagerness of the literature they consult suggests that writers proposing such measures are largely unaware of the content and structure of the ecosystems they recommend be put under siege. It is instructive therefore to proceed to the Great Smoky Mountains National Park, one of the best-studied American reserves, and to reflect briefly on the breakdown of the numbers of known species in each group of organisms. The data are summarized in the table at the end of this chapter. Fifty thousand person-hours of search by research specialists and trained volunteers have yielded records of eighteen thousand two hundred species. The actual number, especially when all suspected but still unrecorded transient species and microorganisms are added, has been estimated to lie between sixty thousand and eighty thousand.

If any of the eighteen thousand two hundred known species seem superfluous to you, please think again. They are just unfamiliar to you, as they are to many of the most devoted scientists as well.

There exist in the Great Smoky Mountains National Park proboscis worms and bristletails and symphylans that might be removed without major consequence to the remaining biota (even here I could easily be wrong), but I am sure that very few of the remainder could be eliminated without the serious decline of some other group of species. To illustrate this principle, consider the consequences of extinguishing any block of five that you might randomly choose from the list in the table—for example, nemerteans, mollusks, annelids, tardigrades, and arachnids. The extirpation of any such block of five would perturb the ecosystem, and even trigger a collapse of the whole.

It follows that no study of a self-sustaining natural ecosystem can be entirely sound without a full biodiversity census of the kind under way in the Great Smoky Mountains National Park. And that is just a beginning. Far more information is needed, including where each species lives, where and when it is active, its life cycle, its population dynamics, its interactions with other species within and outside the ecosystem. And even *within* taxonomic groups, such as all the swallowtail butterflies collectively or all the birds of prey or land snails or orb-weaving spiders and on through the full taxonomic roster, there typically exist great differences among species in basic biology and impact on other organisms.

When as a graduate student I first visited the Great Smoky Mountains National Park, I was fascinated by springtails, part of the group listed in the table as Collembola. These tiny and very elusive creatures possess a lever beneath their bodies, free at one end and attached at the other, allowing it to be opened and closed like a jackknife. When the collembolan is approached by a predator, it releases the free end, and the lever strikes the ground. Milligram for milligram, the strike is one of the most powerful locomotory forces in the animal world. It carries the collembolan high into the air and forward as far as, for humans, would be the equivalent length of a football field.

In the larger picture of natural history, however, this maneuver is only a part of the story. The evolution of predator and prey, it has often been said, is an arms race. It happens that certain kinds of ants have developed ways to best the collembolans at their own game. They use one or the other of two techniques to defeat the high jumps of the prey. One is to put so many huntresses into the field that when a collembolan jumps away from one, it lands close to another.

The other adaptation is one of the most precise and exquisite hunting techniques in the entire animal world. I've studied it in several species of ants belonging to a taxonomic group called the Dacetini. The ant releases a scent from a mass of tissue on its waist that attracts collembolans. Sensing the approach of a collembolan, the ant freezes. Then, guided by odor receptors on the tip of its two antennae waved side by side, it moves very slowly toward the prey. Its long-toothed mandibles are spread widely, in some species more than 180 degrees, locked into position, poised to strike. Two long trigger hairs extend forward, out of the way of the open mandibles. When one or both touch the collembolan, the mandibles snap shut—instantly, faster than the eye can follow, faster than the collembolan can spring away. The mandibles impale the prey with the long teeth lining their inner surface. The collembolan then releases its lever instantly afterward, to no avail—predator and prey remain locked together as they travel through the air.

One day recently as I entered the park, I lifted a piece of bark from a fallen log (permission from a ranger) and saw three tiny symphylans, another group of reclusive insectlike creatures on which the collembolan-hunting ants sometimes feed. These creatures, belonging to a special group of symphylans called japygids, have a pair of pincers on their rear ends. Although there are many species of these worldwide, very little is known about any part of their biology. What is their preferred food, what is their life cycle, why are these pincers in such an unusual position? I have no idea. Nor can I, or any other biologist, guess what would happen if all were to disappear. What I thought at the moment was that if I had another lifetime to live, I could easily devote it to symphylans.

Consider in the same light the different kinds of flies alone you would capture in a field on a summer day by simply sweeping a butterfly net back and forth through the vegetation. (Try it; you'll be amazed.) Then ponder further, if you will, that the species are variously specialized to feed on particular kinds of fruit, or pollen, or fungi, or scat, or dead bodies, or, if you let them, your own fresh blood. Some flies are parasites on other insects—and not just any insects, but one to several species on which they are specialized out of thousands available. When as a teenager I had this epiphany, and I almost became a dipterist, an entomologist who studies flies. I was mesmerized by delicate little flies of the family Dolichopodidae, glistening all over in metallic blues and greens as they displayed themselves on plant leaves in the summer sunshine. How many species were there? Why were they dancing with the maximum effect to be seen, at least to me? Where and how did they spend their lives as larvae? Then I was distracted by ants. Although I was in northern Alabama at the time, far from the tropics, I discovered a colony of army ants crossing our backyard. They were a native species, a miniature version of the army ants that rampage through the forests of Central and South America. I followed their swift-moving column as they crossed a neighbor's yard, then poured across a paved street, and into a patch of woodland. At the tail end of the procession I saw parasitic silverfish and other insect followers. My beautiful little dolichopodid flies could offer no competition to this spectacle, and I resolved to specialize on the study of ants. I had no idea then of the breadth and endless beauty of the world I was entering.

Each ecosystem, be it a pond, meadow, coral reef, or something else out of thousands that can be found around the world, is a web of specialized organisms braided and woven together. The species, each a freely interbreeding population of individuals, interact with a set of the other species in the ecosystem either strongly or weakly or not at all. Where in most ecosystems even the identities of most of the species are unknown, how are biologists to define the many processes of their interactions? How can we predict changes in the ecosystem if some resident species vanish while other, previously absent species invade? At best we have partial data, working off hints, tweaking everything with guesses.

Those of us who have actually analyzed ecosystems to the species level in the field have managed to make serious progress only on the most restricted and elementary among them, and then for only a fraction of their faunas and floras. With a broad brush we have thus depicted mangrove islets, small ponds, tide pools, and Antarctic dry-rock oases. We have learned a few principles from these miniature habitats about the process of colonization, and discovered surprising facts about the relation of predation and of colonization to the amount at equilibrium of biodiversity. We can tell you a bit about the impact of seasonal and climatic variation and the outcome of some kinds of human disturbance. Yet we are forced to admit that the discipline of ecosystem analysis is no more advanced than that of, say, physiology and biochemistry in the early twentieth century, before the revolutions of molecular genetics and cell biology.

What does knowledge of how nature works tell us about conservation and the Anthropocene? This much is clear. To save biodiversity, it is necessary to obey the precautionary principle in the treatment of Earth's natural ecosystems, and to do so strictly. Hold fast until we, scientists and the public alike, know much more about them. Proceed carefully—study, discuss, plan. Give the rest of Earth's life a chance. Avoid nostrums and careless talk about quick fixes, especially those that threaten to harm the natural world beyond return.

Great Smoky Mountains National Park Species Tally*

TAXON	OLD RECORDS (prior to All Taxa Biodiversity Inventory)	NEW TO PARK (since All Taxa Biodiversity Inventory began)	NEW TO SCIENCE	TOTAL RECORDS
Microbes				
Bacteria	0	206	270	476
Archaea	0	0	44	44
Microsporidia	0	3	5	8
Protists	1	41	2	44
Viruses	0	17	7	24
Slime molds	128	143	18	289
Plants				
vascular	1,598	116	0	1,714
non-vascular (mosses, etc.)	463	11	0	474
Algae	358	566	78	1,002
Fungi	2,157	583	58	2,798
Lichens	344	435	32	811
Cnidaria (jellyfish, hydra)	0	3	0	3
Platyhelminthes (flatworms)	6	30	1	37
Bryozoa (moss animals)	0	1	0	1
Acanthocephala (spiny-headed worms)	0	1	0	1
Nematomorpha (horsehair worms)	1	3	0	4
Nematodes (roundworms)	11	69	2	82
Nemertea (ribbon worms)	0	1	0	1

TAXON	OLD RECORDS (prior to All Taxa Biodiversity Inventory)	NEW TO PARK (since All Taxa Biodiversity Inventory began)	NEW TO SCIENCE	TOTAL RECORDS
Mollusks (snails, mussels, etc.)	111	56	6	173
Annelids (aquatic worms, leeches, earthworms)	22	65	5	92
Tardigrades (waterbears)	3	59	18	80
Arachnids				
mites	22	227	32	281
ticks	7	4	0	11
harvestmen	1	21	2	24
spiders	229	256	42	527
scorpions, pseudoscorpions	2	15	0	17
Crustaceans				
crayfish	5	3	3	11
copepods, ostracods, etc.	10	64	26	100
Chilopoda (centipedes)	20	17	0	37
Symphyla (symphylans)	0	0	2	2
Pauropoda (pauropods)	7	25	17	49
Diplopoda (millipedes)	38	29	3	70
Protura (proturans)	11	5	10	26
Collembola (springtails)	64	129	59	252
Diplura (diplurans)	4	5	5	14

continued

TAXON	OLD RECORDS (prior to All Taxa Biodiversity Inventory)	NEW TO PARK (since All Taxa Biodiversity Inventory began)	NEW TO SCIENCE	TOTAL RECORDS
Microcoryphia (jumping bristletails)	1	2	1	4
Thysanura (silverfish)	1	0	0	1
Ephemeroptera (mayflies)	75	51	8	134
Odonata (dragonflies, damselflies)	58	35	0	93
Orthoptera (grasshoppers, crickets, katydids)	65	37	2	104
Other "Orthopteroids" (roaches, mantises, walking sticks)	6	7	0	13
Dermaptera (earwigs)	2	0	0	2
Plecoptera (stoneflies)	70	48	3	121
Isoptera (termites)	0	2	0	2
Hemiptera (true bugs, hoppers)	276	361	3	640
Thysanoptera (thrips)	0	47	0	47
Psocoptera (barklice)	16	52	7	75
Phthiraptera (lice)	8	47	0	55
Coleoptera (beetles)	887	1,580	59	2,526

TAXON	OLD RECORDS (prior to All Taxa Biodiversity Inventory)	NEW TO PARK (since All Taxa Biodiversity Inventory began)	NEW TO SCIENCE	TOTAL RECORDS
Neuroptera (lacewings, antlions, etc.)	12	38	0	50
Hymenoptera (bees, ants, etc.)	245	574	21	840
Trichoptera (caddisflies)	153	82	4	239
Lepidoptera (butterflies, moths, skippers)	891	944	36	1,871
Siphonaptera (fleas)	17	9	1	27
Mecoptera (scorpionflies)	15	2	1	18
Diptera (flies)	599	651	38	1,288
Vertebrates				
fish	70	6	0	76
amphibians	41	2	0	43
reptiles	38	2	0	40
birds	237	10	0	247
mammals	64	1	0	65
TOTALS:	9,470	7,799	931	18,200

*Source: Becky Nichols, Entomologist, Great Smoky Mountains National Park (as of March 2014)

Ivory-billed woodpecker on a willow oak. Mark Catesby, 1729.

II

THE LORD GOD SPECIES

It may seem to you that the study of biodiversity is conducted in a different culture from that of conventional biology. But there are parallels. The cell and brain are like ecosystems—the equivalent of rain forests, or savannas, or coral reefs, or alpine meadows. The location and functions of their various parts must first be discovered and described, then related to create a picture of the whole. But research on organ systems takes place mostly in the confines of a laboratory. Even a square meter of table space may be enough to make a great scientific discovery. In contrast, research on biodiversity—call it biodiversity studies or scientific natural history or evolutionary biology—ranges widely over the entire surface of the planet.

There exist two kinds of scientists. The first go into science in order to make a living. The second do the reverse: they find a way to make a living in order to go into science. Virtually all scientific naturalists of my acquaintance belong to the second group. They are among the hardest-working yet least competitive of all scientists.

They are also among the lowest paid and least garlanded with honors, so there is little incentive to work except for the science. When naturalists meet, they seldom gossip about colleagues not present. Instead they gossip about discoveries, and breaking news. (*"I hear Peter fell down a ravine in El Salvador. Do you know if he's okay?"*)

With rare exceptions, nobody keeps a secret. Quite the opposite. The ethos is to spread the word. If you eavesdrop, you may catch fragments that sound something like: *"Did you hear about that crazy symbiotic katydid Barbara found in an oropendola nest? In Suriname, I think."* Or, *"Bob got to work on his lichens in the Altai, and the Russians gave him permission to camp there at mid-elevation for six months. Boy, I'd be happy to get in there and collect bark beetles for just one week! That's still virgin territory. At least for bark beetles."*

Here's a real example.

Ed Wilson to Ben Raines of the *Mobile Press Register* at the floodplain forest of the Mobile-Tensaw Delta in south Alabama, April 27, 2014: *"I've heard there are jaguarundi in here. That would be a big addition to the mammal fauna."* (Jaguarundis are a species of rare, elusive wildcats known to range naturally from the American tropics north and east to Texas. An introduced population in Florida and on the central Gulf Coast is possible.)

Raines: *"Oh, yeah? Have you seen a photograph of one?"*

Wilson: *"No. I suppose you're going to tell me that nobody has a photograph or skin. But the rumor is at least intriguing."*

Raines: *"Well, I know for a fact that cougars are here, but you almost never see one. So maybe jaguarundis will show up someday, too. We can hope."*

The reason, I believe, for the easy camaraderie is the virtually unlimited number of discoveries that await the trained naturalist with even a little ambition, a discerning eye, and an endless sup-ply of mosquito repellant. The average reward in discoveries per week for hard work is extraordinarily high. It's like getting a bite every time you drop a baited hook into a Louisiana catfish pond. Take three breaths and pull one up. Almost every field trip into a wildland or excursion through a museum collection yields a worth-while discovery in scientific natural history. Provided, of course, you know the species you're dealing with.

Natural historians, like other scientists, have dreams of great dis-coveries, of phenomena unimagined or at best elusive shadows until the precious *aha* moment. We have our holy grails also, the most familiar of which are the missing links of evolution—dinosaurs that morphed into birds, for example, or lungfish into amphibians, or apes into humans.

Equally exciting is the rediscovery of a species thought extinct. Which brings me to one of my favorite fantasies. *I'm paddling a canoe through the back channels of another coastal floodplain forest, this one surrounding the Choctawhatchee River, which crosses the Florida Panhan-dle and flows into the Gulf of Mexico. I hear an eerily familiar birdcall,* peet-peet, *and the double strike of bill against hardwood. I think,* That sounds like . . . but, no, it can't be. *But, yes, it can. Why not?*

In the dream a pair of large woodpeckers swoop in and land noisily on a cypress trunk about twenty feet straight ahead. Then comes another sharp double knock as one of the birds chisels out a beetle grub.

Binoculars up. There is no mistake. There are the black bodies and tails, flashing white primaries of the wings, bright red crest of the male. Ivory-bills! But, I think, Impossible! Maybe I remembered the field guide wrong. *No, I have to believe my own eyes. The last verified sighting of an ivory-billed woodpecker was in the Singer Tract of Louisiana in 1944. In the fantasy I know that. I also remember that a sighting in the Big Woods swamp of Arkansas in 2004 was a mistake; the bird was a pileated woodpecker, a common look-alike of the ivory-bill. And I recall that later a team of birders reported signs of an ivory-bill population in the floodplain forest of Florida's Choctawhatchee River, but that didn't pan out, either. Hasn't yet, anyway.*

Returning from the daydream to reality, I made a trip up the Choctawhatchee several years after the team of birders who'd reported signs of ivory-bills. It was a thrill—even without ivory-bills. The Choctawhatchee is a classic coastal river of the region: dense subtropical floodplain forest, inlets that lead nowhere, life teeming everywhere, turtles of several species called scooters and sliders slipping off fallen tree limbs and out of sight as you approach. Alligators, occasionally big ones, crashing into the water farther upriver.

I nurtured the faint forlorn hope I might see an ivory-bill myself. My companion and host MC Davis, a leading landowner and conservationist of the Gulf Coast who knew the region intimately, was entirely skeptical. He said, "Look, if I wanted to know for sure about ivory-bills, I'd go up the river a ways, talk to some of the swamp rats who live along the river there, and offer a pretty good reward for proof of an ivory-bill population, and I don't mean a dead bird, either."

Anyway, one day I was having breakfast with a group of naturalists in MC Davis's renovated barn, preparing for a day in the field, and the subject of possible ivory-bills in the nearby Choctawhatchee floodplain came up. There was a general thumbs-down—nobody likes to appear gullible—when one of the group said he wanted to play a tape of a birdcall. We hear *peet-peet, peet-peet*. An experienced birder present said quietly, "That's an ivory-bill." It might have been, but I'm a realist. I suspect the tape had been made sixty years previously in the Singer Tract of Louisiana. One of the most spectacular birds of America, the equal in legend of the whooping crane and California condors, the ivory-bill had been driven to extinction by the clear-cutting of cypress and other large trees, on which its existence depended. The Singer Tract was its final refuge, and when that, too, was logged, the species disappeared.

So, to return a moment to the fantasy, imagine what you yourself might say first thing on making this historic discovery. You know this much: if you had lived in the southern United States a hundred years ago, when the ivory-bill was already rare but still occasionally seen, and you had previously never seen one, you might have responded with what was then a common expression: "Lord God, what is that?" Anyone who has been startled, as I have (in Costa Rica), by a pair of large woodpeckers descending suddenly out of nowhere ten feet away at head level will understand that exclamation.

And thus it came to pass that a common name for the ivory-bill in the old days was the Lord God bird.

I'm telling this story not to get you down to the Choctawhatchee but to help convey to you the passion of the naturalist. We each commonly make a discovery of at least one species, whether new to science or rescued from oblivion, or just rare and unexpected, in what may be called a Lord God moment. It could come in the field from anywhere and at any time. Depending on one's expertise, the creature before you could be a Lord God salamander, for example, or a Lord God butterfly, or a Lord God spider, or even, descending the scale all the way down through the vast array of biodiversity, a Lord God virus. Every species still alive is precious to us. Naturalists live for our Lord God moment. We want to save the experience for all generations to come.

G H Ford

W West imp

1. Thamnocenchris aurifer

*In Central America, a venomous palm viper (*Thamnocentris [Bothriechis] aurifer*) captures a black-eyed leaf frog (*Agalychnis moreletii*). Proceedings of the Zoological Society of London, 1848–1860.*

12

THE UNKNOWN WEBS OF LIFE

I t should be obvious to scientists and the general public alike that in order to save biodiversity it is necessary to understand how species interact with one another to form ecosystems. Yet so poor is our knowledge of the interactions that ecosystems studies remain an underdeveloped science, with few answers to pass on that solve even the simplest problems of conservation.

I feel obligated, as a scientist with research experience in both field and theoretical research on ecosystems, to reemphasize in some detail the weakness and inadequacy persisting in this important branch of ecology. Conventional ecosystem studies fall especially short in accounting for the ways in which species interact with one another. Sophisticated mathematical models are available, of course. When data are few, the models are easy to construct—too easy.

Don't get me wrong. Ecological studies at every level are necessary, and attractive. For young scientists with strong mathematical backgrounds they offer a bright future, even *aha* moments. Nevertheless, as a contemporary field of pure scientific study they are even worse off than economics. In parallel manner to this marginal science, the shortcoming is in the database of the identity and natural history of the species making up the ecosystems—much the same as for innate and learned behavior of individual people in economics. Add to that the ubiquitous nonlinearities that twist and turn like escaping eels when you put together the actions of real players. Overall, theorists have not been able to grasp the near-bottomless complexity of the real world, where not just two or three but large numbers of players, species or humans, usually to an indeterminate number, are engaged.

In a few cases ecologists can mine databases for partial correlation that reveal the causes of environmental change. One is the increase of bark beetles in coniferous forests caused by global warming and followed by more frequent forest fires, a general principle. Another is the increase of average niche breadth of species as the number in the ecosystem declines, as seen in north temperate and arctic ecosystems. Individual species on average occur in a greater number of habitats. They consume a greater variety of food. Still another global trend is the rise in diversity of lichens, conifers, and aphids northward around the world, parallel to the decline of orchids, butterflies, and reptiles over the same geographical gradient. As a rule, however, key environmental factors of biodiversity change have been adduced only when the ecosystems studied were very small and their biodiversity relatively simple.

Ecology, like all scientific disciplines, is made up of subjects that are best mastered from the bottom up. First you discover certain phenomena, or infer that such phenomena exist, then you interpret their cause and effect, fitting what you have in hand to whatever explanation seems relevant. With the existing explanation, using common sense or inspiration, you devise possible studies (hypotheses), preferably multiple explanations that compete with one another. You are thereby guided in the search for more data and improved theory that might or might not shed more light on each phenomenon and the pattern of which they are a part. And if you remain unable to explain the whole, at least you can open new avenues of research.

Scientific research is thereby seldom straightforward. It rarely takes major leaps straight to the top. It moves obliquely, pressing forward at angles, reformulating, twisting, filling the subject out, waiting, looking around, describing parts more exactly, describing causal lineages more firmly. Then, like a crack in a cave wall, a guiding beam of light comes through.

That is the way almost all successful science is done. And it is obviously not being done enough in ecology. The data needed to advance studies of the structure and function of ecosystems do not in most cases exist. Let us ask of ecologists, over and again, how can we understand the deep principles of sustainability of a forest or river if we still do not know even the identity of most of the insects, nematodes, and other small animals that run the finely tuned engines of the energy and materials cycles? Turning to the sea, what are we to make of the superabundance of bacteriophage virus predators, which first came to light only in 2013? And how will marine ecosystems be reliably understood when ultra-small organisms, the Picozoa, which might be a key part of the colloid-consuming "dark matter" of the oceans, were first described anatomically and ranked as a new phylum—also only in 2013?

Let me put this matter of ecology's shortfall as science another way. Every scientific discipline must go through a natural history period before it can be synthesized into something resembling mature theory. The scientific natural history lacking in most domains of ecology is the identity and biology of the species that compose biodiversity. At least two-thirds of the species on Earth remain unknown and unnamed, and of the one-third known, fewer than one in a thousand have been subject to intensive biological research. In the same way that physiology and medicine could not have advanced (nor been properly taught) without a solid knowledge of the organs and tissues of the human body, serious future advance of ecosystems analysis cannot be expected to emerge without a solid knowledge of the species composing one ecosystem at a time.

Writers and spokespeople favoring Anthropocene philosophy are focused on ecosystems and by training seem innocent of the nature and meaning of biodiversity at the species level. Researchers in species-level biology are the equivalent of neurobiologists in their finely detailed study of the brain, while those Anthropocene enthusiasts who see species as interchangeable parts that fill up ecosystems are little more than nineteenth century phrenologists, who studied mind by shape of the skull.

It follows that most of the necessary work on ecosystems immediately ahead lies in the study of biodiversity at the level of species. The exploration of biodiversity starts, as it always has, with taxonomy. Taxonomists discover species, learn to identify them by differences in anatomy, DNA, behavior, habitat, and other biological traits. All of this information has practical value. Suppose that a new kind of fruit fly introduced from a still-unknown location threatens the western alfalfa croplands. What is the name and identity of this invader? Where did it come from? What are its parasites and other natural enemies in the homeland? What else can be taken from its known biology to help control it? It would not be wise to wait for each such emergency in order to begin the needed research from the ground up. Bear in mind that the number of such invasive species is growing exponentially everywhere in the world. A few in the rising tide are potential pests. Another small fraction are disease-causing microbes. Yet another fraction are insects and other organisms that carry the pathogenic microbes from one human or domestic animal to another.

Consider a second kind of problem becoming increasingly urgent in conservation. In 2014, officials of the oil palm industry proposed to cut down and convert half of Borneo's rain forest, leaving half as a conservation reserve. What effect would this massive destruction have on Borneo's biodiversity? Would all of the island's species survive in the trimmed-down reserve? Or would it be closer to 80 percent, or to half? And in the process, how many species found nowhere else in the world would fall to the chain saw? Previous experience with widespread conversion of natural environments suggests that the loss would be less than half, but still in the range of 10 or 20 percent, with many of the species found only in the destroyed half lost forever, or doomed to early extinction, before they are even known to science.

Another problem is the claim by Anthropocene enthusiasts that there are no more wildernesses, that Earth is already a used planet, and unfettered nature is dead or dying. The time has come, they say, to bring people more fully into the picture, mingling people and wild species in a manner that benefits the two symbiotically. So just how many species, and how much nature, would survive? Anthropocene supporters have no idea, and qualified scientists are struggling to find out.

I've stressed species as the unit in the hierarchy of biodiversity (ecosystem, species, gene) that can and must be studied most thoroughly with the practical aim of saving the whole. What needs to be done? The history of each species can be considered an epic. The entire biology of one species can consume the lifetime of a scientist. Even with a hundred scientists focused on the species, our knowledge of it remains incomplete. There is the niche, where the species lives intimately with other species—prey, predator, inside and outside symbionts, engineers of soil and vegetation. Therein, we can find no species that lives alone. When we allow one species to die, we erase the web of relationships it maintained in life, with consequences that scientists seldom understand. In displacing wildness, our actions are ignorant and permanently destructive. We break many threads, and change the ecosystem in ways still impossible to predict. As the pioneer environmentalist Barry Commoner observed as his Second Law of Ecology, "You cannot do just one thing."

The major binding ties of ecosystems are food webs. Insects, as taught in Ecology 101, prey on plants, birds prey on insects, plants feed birds with seeds and fruit containing seeds, birds spread seeds when they defecate, promoting the growth of plants. With predator-prey and symbiotic links of this simplicity, in a small ecosystem of a few species, it is possible to build mathematical models that predict population cycles, dispersal, and the likelihood of the persistence of the species. But are the models true?

It is highly unlikely they will prove to be true except within very broad limits. Ecologists are well aware that webs of such simplicity are rare in nature, for reasons I have stressed. The real world, when explored as scientific natural history, reveals relationships among species that are usually surprising and often bizarre, far from ordinary human experience. I offer the following examples to make my point.

VAMPIRE HUNTERS. The five thousand species of jumping spiders known worldwide are mostly short-legged, chunky in build, and hairy. They do not build webs, but instead prowl over the ground and vegetation using their large eyes to search for prey. When one is seen, the spider approaches and pounces on it, catlike. The spiders vary in size such that if the smallest were the size of a domestic cat, the largest would be as big as a lion. If, while relaxing in your backyard reading a newspaper, you see a small chubby spider appear on the paper and start to walk over it in short zigzag runs, your visitor is almost certainly a jumping spider. They are specialized on the prey they hunt. Some prefer ants, others hunt spiders of other species. *Evarcha culicivora* of East Africa prefers mosquitoes—and not just any mosquitoes, but vampirish females that have recently fed on the blood of humans and other vertebrates. The *Evarcha* are especially common around houses, and in their small way contribute to malaria control. (*Culicivora* means "mosquito eater.")

ZOMBIE MASTERS. As a second example of evolutionary idiosyncrasy, consider the ingenuity of parasites. The larvae (caterpillars) of the European gypsy moth avoid birds and other predators during the day by hiding beneath the bark of trees. When darkness falls they climb back up to the canopy for their evening meal of leaves. The daytime behavior is reversed, however, when the caterpillars are infected with a virus (specifically, a nucleopolyhedrovirus) that occasionally sweeps with deadly effectiveness through the gypsy moth population. The virus induces a change in the brain that causes the caterpillar to climb in daylight to the tops of the trees. There its body liquefies and releases a cloud of the virus that infects other caterpillars—especially when aided by rainfall. A similar zombie-like control is exercised by cordyceps fungi, which infect ants as they forage on vegetation. The ants, in their last dying act, clamp their jaws onto a leaf vein, an act that secures each corpse in place. The fungus then grows out of the body and disperses its spores into the air to fall on other ants.

SWINDLERS. The exactitude of species evolution often includes trickery. The natural world is rife with plants and animals that emit false clues in order to complete their life cycles. One of the most elaborate is employed by a species of blister beetle (*Moloe franciscanus*) found in the southwestern United States. It uses a series of tricks to steal the resources of a solitary bee (*Habropoda pallida*) common in the same habitats. The blister beetle female first lays eggs at the base of plants regularly visited by the bee for pollen and nectar. Upon hatching, the beetle larvae climb up on the plant and cluster together to form a small spherical mass. The little swindlers then release an odor with substances used by female bees to attract males of their own species. The beetle larvae pile onto the back of the defrauded male. When the male bee contacts a real female bee of his own species and mates, the beetle larvae leave him and gather on the back of the female. They ride her until she returns to her nest, whereupon they climb off and eat the pollen and nectar stored there, as well as the egg she has laid.

Orchids are the master swindlers of the plant world. An entire encyclopedia could be written about their duplicity. Comprising seventeen thousand species, making them the largest of all flowering plant families, they use a wide variety of tricks that induce insects to transport pollen to other plants of their own species. Some, for example, have flowers that resemble the females of certain species of wasps. When the males alight to copulate, all they receive are sticky masses of orchid pollen that cling to their bodies. Other orchid species draw in males by releasing the scent of a female. At least one species emits the odor of angry bees, which attracts wasps of a kind that prey on these insects. The wasps, with the fake-bee orchid pollen masses sticking to them, end up pollinating orchids instead of searching for mates of their own species or gathering food.

SLAVE MAKERS. The social insects provide many bizarre examples of extreme adaptation. In the north temperate zone high drama has been played out amid deceit in the millions of years of ant slavery. There are many species that raid the colonies of other ant species and, after driving off the defending adults, kidnap (antnap?) the immature offspring in their helpless pupal stage. The raiders then raise the captives (allow them to emerge from the pupae) as slaves to augment their own worker force. In extreme cases the slave makers, like Spartan warriors of old, depend entirely on the quotidian labor of the slaves. The deception is based on a universal trait in ants: after workers emerge from the pupal stage as adults, they fixate during the first several days on the odor of the colony around them. From that time until they die, the captives regard the slave makers as sisters, not oppressors as they truly are.

The slave makers are ferocious warriors. Some are equipped with powerful sickle-shaped mandibles used to kill or cripple the defenders of the colonies they target. One species I studied achieves the same result with "propaganda" substances. During attacks the raiders release large quantities of a chemical that acts as an alarm signal for the defenders. The besieged workers are thrown into a panic, in the way humans might be if suddenly blasted with the scream of fire alarms set at high volume all around them. The raiders themselves are not disturbed; they are instead attracted to the odor of their own pheromone.

What would slave makers do if made to live without captive labor? One day I decided to find out by removing all the slaves from slave-maker colonies I kept in my laboratory. The warriors, left on their own and with no previous experience, began to attempt the tasks previously left to the slaves. But they performed badly. They attended the young in a slovenly manner, picking the larvae and pupae up, carrying them about for a while, then dropping them in the wrong places. They were entirely unable to bring food into the nest, even when I placed morsels next to the entrance.

What would happen if slave makers disappeared from the many ecosystems their various species inhabit? I have no idea—the impact of parasites is a field all of its own.

GIANT KILLERS. Naturalists are often startled when they first study a previously unknown species. When reflecting on the universal phenomenon of predation among animals, for example, naturalists tend to assume that predator and prey are roughly the same size, or else the predator is much larger. There are exceptions, however. Birds peck insects. A pack of wolves can bring down a moose in deep snow, and a pride of lions once in a while can even disable an elephant. Still, the weight difference between predator and prey is usually no more than tenfold. A number of exceptions, however, are found among the ants. None exceeds the drama of the South American species *Azteca andreae*, which lives in *Cecropia obtusa*, a broad-leafed tree in the rain forest of South America. As many as eight thousand workers line up side by side on the lower margins of a single *Cecropia* leaf. There they wait, with their mandibles held open, ready for immediate action. When an insect lands on the leaf, the ambush team rushes it from all sides, and, working together, pin and spread-eagle it. By this means insect prey of almost any size can be caught. One, pulled to safety and measured, weighed 13,350 times a single worker ant. This amazing innovation, the equivalent of mammoth hunters among early humans, appears to be rare in nature.

I've chosen the foregoing examples of species interactions, all very peculiar in nature, with a purpose. First, to hold your attention, of course; insects are not everyone's favorite animals. But they also illustrate an important principle of ecosystems studies. Imagine for yourself any niche that might have evolved on this planet within the realm of physical possibility. (No animal can run a hundred miles an hour, for example, or digest iron ore.) Somewhere among the millions of species on Earth one or more will probably occupy the niche you chose. To extend this principle from species to ecosystems, there is no better way than the exclamation that closes Charles Darwin's *Origin of Species*:

> It is interesting to contemplate an entangled bank, clothed with many plants of many kinds, with birds singing on the bushes, with various insects flitting about, and with worms crawling through the damp earth, and to reflect that these elaborately constructed forms, so different from each other and dependent on each other in so complex a manner, have all been produced by laws acting around us.

To those who think nature consists primarily of plants and large vertebrate animals, I say look about you at the little things that run the earth. To those who believe they can fathom the working of ecosystems with mathematical models of a handful of species, I say you live in a dream world. And to those who believe that a damaged ecosystem will heal itself or can be safely restored by replacing original native species with functional alien equivalents, I say think again before you cause damage. Just as successful medicine depends upon a knowledge of anatomy and physiology, conservation science depends upon a knowledge of taxonomy and natural history.

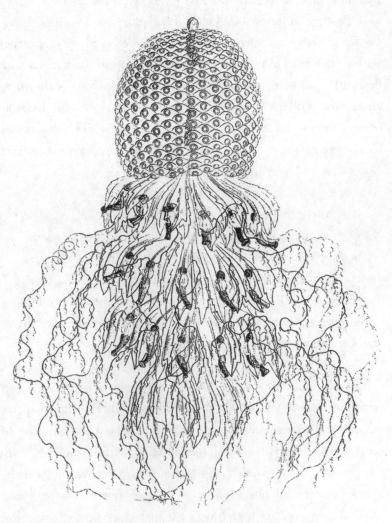

The siphonophore Forskalia tholoides *(colonial "jellyfish"),*
modified from Ernst Haeckel, 1873–1876.

13

THE WHOLLY DIFFERENT
AQUEOUS WORLD

Earth is invested with two different living worlds, two radically different arrays of ecosystems. The basic hierarchies of biodiversity are the same, from ecosystems to species to genes, and they face the same threat of obliteration. But all the rest is different.

To explain I propose to take you on a brief journey. Come with me down to the edge of the sea. Look out to an environment so removed from the land and sky as to seem on another planet. If you were to submerge yourself into it physically without a life-support system, you would die in less than ten minutes. There are vast areas on the bottom that have, in fact, never been visited by humans, much less closely viewed.

Most of the marine world has hitherto lived outside the events of the Anthropocene, but that is changing rapidly in the early years of the twenty-first century. Humans have assaulted even the farthest and deepest ocean waters, especially wherever food and other resources can be profitably extracted. Our ecological footprint is growing: the waters are warming and acidifying; coral reefs are fading to oblivion and in some places are being terminally dynamited. The high seas are routinely overfished. The benthos is turned into barren mud by bottom trawling, and dead zones are smeared out over the bottom from polluted river deltas.

Yet most of marine biodiversity persists. The populations of many of the species have been reduced and their geographical ranges have grown smaller, but few have yet been driven to final extinction. The sea still has places to witness the manner in which species fit together into healthy ecosystems. Most of it, as I'll now explain, is even intact and still in an early period of exploration.

Let's start at the beach. Stand, if you will, on the wet sand of the surf zone at low tide, letting longer waves work around and over your feet, pulling sand out from under them and starting to accumulate it on top. Now think biology. The surf may at first seem lifeless, composed of water and soil washed clean. The opposite is true. It is home to a legion of invertebrate animals unique to this habitat. They range in size from bullet-shaped burrowing shrimp ("sand fleas") the size of a thumb down to the vast majority, which are barely visible if at all to the naked eye.

The strangeness of the meiofauna, as it is called (the Greek word *meior* means "less"), especially given the simplicity of its habitat, is not just in the species it contains but in the many higher taxonomic categories represented by the species. If on the land you walk up to the edge of a forest and census its animal biodiversity carefully, you will likely find representatives of the following seven phyla: Chordata (birds, mammals, amphibians), Arthropoda (insects, spiders, mites, millipedes, centipedes, crustaceans), Mollusca (snails, slugs), Annelida (earthworms), Nematoda (roundworms), Tardigrada (bear animalcules), and Rotifera (rotifers). Among the grains of sand in the surf zone you will in time find twice this number of phyla (causing you perhaps to start murmuring over and again, "Lord God"), including entoprocts, gastrotrichs, gnathostomulids, kinorhynchs, nematodes, nemerteans, priapulids, sipunculans, and tardigrades. These teem among the more familiar mollusks, polychaete worms, rotifers, and crustaceans. A common body form of meiofaunal animals is wormlike, allowing them to move quickly through the tightly packed grains of sand. They slither to feed, slither to avoid being eaten, slither to mate and reproduce.

Research on the meiofauna and its place in the shoreline ecosystems of the world is in an early stage. The multiple ways its species interact are still largely unstudied. Yet these strange denizens of one of Earth's most physically dynamic ecosystems are important parts of the biosphere. And there are more of them than you might imagine. Although the meiofauna may live within a strip of habitat only a kilometer wide, the total length of Earth's shoreline is 573,000 kilometers, almost exactly the distance from Earth to the moon. If the meiofauna realm is taken to be this length and the average width guessed to be a kilometer, its total area would be about the same as that of Germany.

Let's pass on by the already familiar offshore coral reefs, often called the rain forests of the sea for their immense structural complexity and biodiversity, and travel out above open water, then drop down to skim the surface. Here you will find the pleuston, an entirely different habitat, unfamiliar even to many marine biologists. Where air touches the sea are organisms specialized to live on or just beneath the layer formed by surface tension. Although sparsely distributed, they are everywhere present in all of the oceans. They live on floating islands mostly of animal origin, which range in size from the corpses of fish and seabirds down to fragments of algae and mucus almost too small to be seen with the naked eye. Each fragment is home to a community of living organisms. The inhabitants always include bacteria of multiple species and probably also archaea, which resemble bacteria but are far removed from them in DNA. Like plants and animals newly arrived on a conventional oceanic island, they flourish and multiply until they have used up all of the available nutrients.

All across the oceans and inland seas, bacteria and archaea ride the surface detritus but also float and swim in the waters below. Those ranging freely obtain material and energy from photosynthesis. The overall result is this: no matter how crystalline-clear the water appears, it teems with life.

Because I am an entomologist by specialty, I take a special interest in marine insects. Because their millions of species and the roiling biomass they compose dominate animal life on the land, it is of considerable interest to know how many inhabit the marine environments. The answer is virtually none, and therein lies an absorbing scientific mystery. In my own research of islands, I've discovered caterpillars living in the submerged prop roots of red mangroves, the most seaward and globally widespread among the mangrove species. But this near-terrestrial habitat is a very far distance from the coral garden and blue water farther out. There are no insects in that world except on the surface, and these odd voyagers are extremely rare. Few marine biologists have even seen one alive. I have not. Linnaeus and Darwin were completely unaware of their existence. All known to exist are water striders, a kind of "true bug" (insect order Hemiptera) common in freshwater streams, ponds, and lakes, where they skip over the surface on long, water-repellent legs. Freshwater water striders prey on other insects, such as mosquito larvae, which themselves live on or close to the surface. All of the marine water striders belong to the single genus *Halobates*. Only five species in this select group have been found living on the open seas. The prey they hunt is unknown.

The existence of *Halobates* water striders on the open sea is an exception that only deepens the mystery. Insects have been evolving on the land and in ponds and other bodies of freshwater habitats for more than four hundred million years. During that period they have dominated animal life wherever there are plants. Undetermined millions of species have evolved and proliferated during repeated surges of evolution. Yet only *Halobates*, water striders, have managed a foot on the sea. No others, to the best of my knowledge, have been found even in a narrow niche among the thousands occupied by species of other invertebrate animals, such as crustaceans, sea spiders, and polychaete worms.

Conrad Labandeira, a leading paleontologist and expert on insect ecology, has suggested that there are no marine insects because there are no trees or lower leafy vegetation of the kind where on the land insect life flourishes. Perhaps so. But there is plenty of layered vegetation in shallow marine waters, such as the kelp forests along the Pacific coast. Even that habitat has somehow not yielded to invasion. Instead, it abounds with predators, parasites, and scavengers belonging to other kinds of invertebrates.

Surprises of a different kind await in a wholly different fauna in yet another biological realm of the ocean, the deep scattering layer. If you are an open-ocean fisherman not committed to marlin, tuna, and other big-game species, you will encounter the widest variety of fishes by going out at night. As the sunlight fades, a dense concentration of fish, accompanied by squid and crustaceans, ascend from a depth of about 270 to 360 meters, where they have cruised and floated hidden in darkness during the daylight hours. The cover deeper water gives them is not complete, however, because predators swimming at even greater depths still see them silhouetted against the light far above. As a second defense some species use counter-illumination, the production on their undersides of bioluminescence, the light provided by their own tissue or symbiotic bacteria carried inside their bodies. The illumination matches the brightness around them coming from the sun or moon above, rendering the animal less visible.

Obedient to a law of biology, every game played between predator and prey in the deep scattering layer is an evolutionary arms race. Thus at least a few deepwater sharks, together with predatory hatchetfish and bobtail squid, have lifted the game to the next level: they have their own bottom lights that hide their approach to the prey.

One of the surprises in recent marine biology coming from the deep scattering layer, aside from the discovery of the layer itself, has been the encounter of a genuine monster living there. The first megamouth shark was discovered in deep water off Hawaii in only 1976. By 2014, over fifty others had been captured or at least reliably sighted. Although the megamouth reaches a length of at least eighteen feet and weighs up to twelve hundred pounds or more, it is a giant not to be feared. Its enormous mouth paradoxically has small teeth and in any case does not bite; it opens wide and funnels in smaller crustaceans and other plankton, the same method used by the equally harmless manta rays, whale sharks, basking sharks, and baleen whales.

If giants still swim unseen on moonless nights beneath ships across the vastness of the sea, what other surprises await us in the smaller creatures swarming among them? That question is very much on the minds of scientists. In order to find the most extreme and least-known forms of life, they have begun to hunt more thoroughly for ocean microbes, some of which in particular turn out to be the smallest known organisms on Earth.

Beetles that live in and on decaying wood. Alfred Edmund Brehm, 1883–1884.

14

THE INVISIBLE EMPIRE

Early in this century the members of the Explorers Club of New York faced a shrinking supply of unclimbed mountain peaks, unwalked polar ice sheets, and unvisited Amazonian tribes. In 2009 they adapted by adding biodiversity to their purview, which turned out to be a wise choice. The exploration of biodiversity offers scientists and adventurers alike the greatest physical adventures that remain on planet Earth.

Remarkably, the new emphasis on biodiversity includes an auspicious examination of the flora and fauna inhabiting our own bodies. The ability to rapidly sequence the DNA of microorganisms has revealed that each healthy person contains a series of balanced ecosystems composed primarily of bacteria. Like microbes resident in other organisms, they are mostly friendly to their human hosts. Called "mutualistic symbionts" by biologists, they both benefit from the plants or animals with whom they live, and give benefit in return.

Over five hundred such species of bacteria live in the mouth and esophagus of the average human. By forming a well-adapted microbial rain forest, they protect this part of the body from harmful, parasitic species of bacteria. The price of failure in the symbiosis is an invasion of aliens, the buildup of dental plaque, tooth decay, and gum disease.

Farther down in each successive part of the gastrointestinal tract, colonies of other specialized bacteria play crucial roles in digestion and waste disposal. The average number of human cells in the body runs at least into the tens of trillions—one number calculated from multiple estimates is forty trillion. The average number of bacteria in our microbiome, as it has come to be called, is at least ten times higher. To make the point, microbiologists joke that if biological taxonomy were to be based exclusively on the preponderance of DNA within each organism, human beings would be classified as bacteria.

It should not be surprising to learn that the microbiome looms large in medical science and practice. Researchers have come increasingly to investigate the symbionts' role in a wide range of health problems, usually those arising in the gastrointestinal tract but also including obesity, diabetes, proneness to infections, and even some forms of mental illness. The microbiome is an interlocked array of ecosystems whose species need to be kept diverse and in correct balance. In a nutshell, much of future medical practice will become a kind of bacterial gardening.

The gardens growing within human beings and other animals are typical of complex ecosystems everywhere, inside and out. The overall number of kinds of microbiomes dwelling in animals and plants worldwide remains entirely unknown, but it must be enormous. Those inhabiting wood-eating termites, microbiologists have found, are drastically different from the ones carried by carnivorous ants, and so a fortiori to other organisms as diverse as frogs and earthworms. It is obvious that symbiotic microbiology, which encompasses these systems, has emerged as an exciting frontier of science, and will remain so for decades to come.

Microbiology as a science was founded by Antonie van Leeuwenhoek in the late 1600s, when he invented microscopes powerful enough to see bacteria. But it was nearly four hundred years later that biologists, led by the American Carl Woese, first recognized that the archaea, a group of bacteria-like microorganisms, are radically distinct in DNA from bacteria. The discovery challenged the basic arrangement of the Tree of Life and our picture of the early evolution of life. The Tree is a branching diagram that displays the ancestral relationships of species and groups of species one to another, thus tracking the course of evolution that occurred as earlier kinds of organisms gave rise to younger ones. It shows how some species multiply into daughter species across millions of years or, in a few cases, as quickly as thousands of years, while others remain undivided.

Until Woese and his colleagues, the standard classification recognized five kingdoms: Monera (comprising the bacteria and archaea), Protista (paramecia, amoebae, and many other single-celled organisms), Fungi, Plantae, and Animalia. After Woese, three sprawling "domains" of life were left: the bacteria, microbes that lack a distinct nucleus; the newly defined archaea, similar to bacteria in their structure and also lacking a nucleus; and the Eukarya, having a nucleus, and combining all other known life-forms—protists, fungi, animals, algae, and plants. What DNA comparisons revealed was that while the Eukarya are mostly large in size—the property we see and care about almost to the exclusion of everything else— bacteria and archaea rule in numbers and global distribution, just as they have since the origin of life.

Microbes create and deposit minerals. They break down and secrete organic chemicals, and influence plant growth. They are everywhere, collectively prepared to clean toxic waste, capture and collect the energy of sunlight, and combine water with carbon. They rule at the base of the food chain. In short, our planet, as the microbiologist Roberto Kolter once summed up the biosphere, "has been shaped by an invisible world."

Genetic diversity in the microbial world stands well apart from that in the rest of life. Human beings and potatoes are closer in their DNA sequences than the most dissimilar species of bacteria are from one another. Biologists have no idea of the number of species of bacteria on Earth. It could be in the tens of millions, perhaps hundreds of millions. At the present time there is not even an exact definition of bacterial and archaean species. Further complicating the picture is the promiscuity of the individual organisms. Bacterial cells have multiple ways of picking up genes from other cells, whether closely related or not. They accomplish that high-technology feat by collecting DNA fragments from the environment; by filching pieces from retroviruses during DNA transport by these quasi-organisms into the cells of larger organisms that host the viruses; and, finally, by engaging in conjugation, a surprisingly complex operation during which two cells come together and swap similar segments of their own DNA.

Even the geography of bacterial diversity may be fundamentally different from that of plants and animals. In some way or other many kinds of bacteria obey a principle first proposed in 1934 by the pioneering ecologist Lourens Bass Becking: "Everything is everywhere, but the environment selects." That is, a large fraction of genetic forms or something very close to them occurs around the globe, but most lay dormant most of the time. They form a microbial seed bank within which each species begins to multiply only when the environment changes to suit its DNA-coded preference. The sleeping cells awake when presented by the right acidity, the right nutrients, and the right temperature. Seed banks occur throughout the terrestrial and watery beds. Only across great distances, perhaps continental in scale, do genetic differences emerge, resembling species formation at the magnitude conventional in plant and animal species.

As DNA technology has improved, so has the rate of discovery of kinds of bacteria and other microscopic life-forms. Some are hidden in plain sight—abundant and exposed, yet too small to be seen with conventional light microscopy. As a result they are easily overlooked in standard screenings of microbial communities.

The most important such organisms discovered thus far may be the *Prochlorococcus* bacteria. Although first recognized only in 1988, *Prochlorococcus* are far from rare. They are in fact the most abundant organisms in tropical and subtropical seas around the world. They live in waters as deep as two hundred meters and reach local densities of more than a hundred thousand per milliliter. Because the tiny cells are also photosynthetic, living on the energy of sunlight, they account for 20 to 40 percent of the biomass of all photosynthetic organisms present between 40°N and 40°S in the open ocean, and they are responsible for up to half of local net primary production. In warm ocean waters, at least, the invisible support the visible.

Yet—if *Prochlorococcus* along with a second superabundant bacterium, *Pelagibacter*, are the most abundant conventional organisms, might they be the prey of viruses that are even smaller? Experts used to think that such micro-predators are at best relatively rare. In 2013, however, new methods in the fast-moving field of ultramicroscopic research revealed the presence of viruses on an average of billions per liter of seawater. All are bacteriophages (literally, "eaters of bacteria"), of which one, HTVC010P, is the most abundant. Biologists disagree on whether viruses in general are true organisms, since they rely on the molecular machinery of their hosts in order to reproduce. But if HTVC010P is thus classified, it must be considered the most abundant known species on Earth.

There is more. While sunlight-capturing *Prochlorococcus* bacteria and its bacteriophage predators are a large part of the known matter of the ocean, they are likely balanced by a still mostly undiscovered "dark matter" of scavengers and predators. These, too, are undetectable by means of conventional microscopy. Apparently filling the role at least in part is a highly diverse assemblage of picobiliphytes that measure only two- to three-millionths of a meter in greatest diameter. One of the species was studied in sufficient detail in 2013 to recognize it as a new phylum, which researchers named the Picozoa. These almost vanishingly small creatures feed on colloidal fragments, moving jerkily through the water by whipping their flagellae back and forth. The anatomy of the one well-studied species is unique among microorganisms. It is oblong in shape, with its feeding apparatus occupying an entire one-half of the body, leaving all the other organelles packed into the other half.

As scientists have probed downward in the ocean, on past the scattering layer at the edge of light and into the cold black, crushing deep, they have discovered yet another world of fishes, invertebrate animals, and microorganisms. Each species is specialized to survive and reproduce at the particular depth where it has evolved.

Then abruptly comes the bottom, darkly and poetically well named the abyssal benthos. Surely, you might think, there can be very little life in this featureless plain of silt so far below the photosynthesizing lighted waters kilometers above. That would be a mistake. The abyssal benthos teems with life. It is a prime hunting ground where the primary producers are not plants and photosynthesizing bacteria but scavengers. These organisms feast on every carcass, every piece of flesh and bone, every flake and salt-grain-sized particle that makes it past the viper fish, gulper eels, and other hunters cruising the waters above them. The bottom-feeders include unique species of fishes, invertebrates, and bacteria. Together they wait for the decaying manna dropped from above. They are the final scavengers, predators of scavengers, and predators of the predators of scavengers in open water. Food is very scarce, but the ecosystem works because all of its inhabitants have a superb sense of smell. They are able to track extremely faint traces of odor rising from every morsel of food, even in relatively still water.

Consider the fate of a wooden boat that sinks at sea. Soon after it settles on the bottom, larvae of *Teredo* wood borers (Teredinidae) locate it and fasten to its surface. As the later stages grow, these "termites of the sea" feed on the wood, boring tunnels as they proceed. The teredos are not worms, which they superficially resemble, but mollusks genetically closest to barnacles.

Next, visualize a piece of meat that makes it to the bottom. Within minutes it is likely to be discovered and vigorously attacked by deep-sea macrourid fish and jawless, eel-like hagfish. Then come the scavenger invertebrates, and quickly thereafter bacteria, and soon nothing remains but final organic products diffusing outward in the icy still water.

Any scraps of leftover wood and flesh are targets of invertebrates and bacteria, among which are specialized scavengers. If a prize were given for "most bizarre," at least as a human would judge it, my vote would go to the *Osedax* worms, which feed on the lipids in the bones of whales that fall to the ocean floor. The diet is peculiar enough, but the method of feeding strains credulity. Female *Osedax*, which grow to the size of a human finger, have neither mouths nor intestines. They feed instead by penetrating the bones with bodily extensions that contain symbiotic bacteria. These microbial partners metabolize the lipids and share the material and energy produced with their worm hosts. And what of the male *Osedax*? Even stranger. The larva-like adults are only a third of a millimeter long, a dot from a pen in the palm of your hand. More than a hundred live in a colony of parasites beneath the sheath of every female, feeding on yolk of the female's eggs—in other words, their own future brothers and sisters. The niche occupied by the *Osedax* is not as scarce as it may seem; about six hundred thousand whale skeletons have been estimated to lie on Earth's ocean floors.

However, even if *Osedax* whalebone eaters make you pause, let us not stop with them. There is another kind of life even farther down. Under the bottom surface, past the benthic crust, life continues to flourish. The marine organisms there join their counterparts, which occupy deep subterranean layers of soil and rock on land, to form a shroud enveloping the planet called the deep biosphere. The inhabitants are overwhelmingly microbial, comprising large proportions of both bacteria and the bacteria-like archaea. In some marine localities, at least, a million of their cells abound per cubic centimeter at half a kilometer below the benthic surface. If this density proves to be worldwide, it means that the microbes of the marine deep biosphere comprise more than half of all microorganisms on the planet. It would possess the same order of magnitude of biomass (weight of organic material) as all the photosynthesizing plant life on Earth's surface.

If these estimates are even approximately correct, the existence of the deep biosphere requires a fundamental shift in our view of both the microbial species and the ecosystems they compose. At the ocean bottom and just below one meter perhaps, the microbes and sparse invertebrate animals among them remain players in the carbon cycle of the waters and land above. Their energy is drawn from the debris of organisms that were created by solar energy. But in the deep biosphere farther down, this connection weakens and is evidently replaced by chemical energy derived in a different manner: geochemical process, the derivation of energy from nonorganic sources in the soil and rocks. The deepest at which all such microbes have been found is 2.8 kilometers below the Earth's surface, in the walls of the Mponeng gold mine near Johannesburg, South Africa. There, in the absence of light and oxygen, subject to a constant sixty degrees Celsius, lives the newly discovered species *Desulforudis audaxviator*. It exists by reducing sulfate and fixing carbon nitrogen from its surrounding inorganic environment. Given the steady rise of temperature with depth, *Desulforudis audaxviator* is, so far as known, the only species present. Its habitat appears to mark the inner boundary of life on the planet.

What of more complex, multiple-celled life-forms? As I write, one species of nematode worm has been discovered in the lower reaches of the deep biosphere, feeding on microbes. More such invertebrates, representing considerable biodiversity, are likely to be found.

The existence of an independent layer of life invites reflection on the imagined Armageddon, and with it the End of the World. If our species, self-appointed Lords of Earth, accidentally burned the planet's surface to a crisp, or an outsized asteroid were to wipe out all organisms on the surface, life could go on in the planet's deep biosphere. There, its microbes and invertebrate predators would persist, shielded and unconcerned in their dark refugia, drawing their energy and substance from rocks, resistant to heat, perhaps evolving over hundreds of millions of years, eventually to reach the surface and generate a diversity of multiple-celled organisms, thence against heavy odds a human-grade metazoan. Whereby the great cosmic cycle might give intelligence on Earth a second chance.

*A European woodland, with two common snipes (Gallinago gallinago) in the
lower center. Alfred Edmund Brehm, 1883–1884.*

15

THE BEST PLACES IN THE BIOSPHERE

Many who have spent their lives in cities or densely populated rural areas find it easy to imagine a world given over entirely to humanity. In the extreme Anthropocene worldview, it seems more logical to re-engineer what remains of nature to serve people than to protect nature in its original condition. Why, its devotees ask, commit space and resources to a lost cause? The natural world is already badly damaged. Its biodiversity is past the point of no return. Primeval habitats no longer exist. This defeatist opinion should be kept clearly in view. As John Stuart Mill once put the essence of discourse, both teachers and learners alike fall asleep when there is no enemy in the field.

Naturalists and population biologists of course see the world in an entirely different way. A point of no return exists, but it only exists for humanity should we devote too much of the planet's environment to the needs and pleasures of our one species. An Earth packed-wall-to wall with people would be a planetary spaceship, dependent on humanity's future intellect and wisdom for the long-term survival of life. It would not only be disastrous for the rest of life but a high risk to our own long-term survival.

Conservation organizations themselves are not immune to the Anthropocene worldview. A worrisome expression of this turn of thought is expressed in recent annual reports of The Nature Conservancy. TNC has been one of the most admired nongovernmental proponents of nature reserves, acquiring and setting aside millions of acres for permanent conservation. This work will no doubt continue, but the organization appears to have adopted a different mood. What nature can do for people and the economy has moved to center stage, while biodiversity has faded. The wraparound cover of the 2013 annual report, for example, is emblematic of what appears to be this worrisome trend. It is a photograph of a smiling boy on horseback tending a herd of goats in Mongolia. Behind him, stretching to the horizon, is a geometrically flat grassland. The total biodiversity thus comprises four species of organisms: a human, two domestic animals, and one plant. Almost every photograph and block of photographs in the text features people, their habitations, and their domestic animals. One portrays elephants, another penguins, a third sandhill cranes, and a fourth salmon fillets hanging in an Alaskan smokehouse, their usefulness to people permitting no doubt.

In contrast, the view of experienced naturalists and conservation biologists is focused on the two million other known species on Earth, and the more than six million others thought still undiscovered. Because it is clear that a healthy biosphere is good for the economy, we trust that the public and business and political leaders among them will join us and come to value the living world as an independent moral imperative that also happens to be vital for human welfare.

The clear lesson of biodiversity research is that the diversity of species, arrayed in countless natural ecosystems on the land and in the sea, is under threat. Those who have studied the database most carefully agree that human activity, which has raised the species extinction rate a thousand times over its prehuman level, threatens to extinguish or bring to the brink of extinction half of the species still surviving into this century. Yet there remain scattered around the world many reservoirs of Earth's biodiversity, from a few acres in area to authentic original wildernesses with areas in excess of many thousands of square kilometers. Almost all of these last domains of the natural living environment are under some degree of threat or other, but they can be saved for future generations if those alive today have the will to act on their behalf.

To document this important point about the promise of an expanded global conservation, I wrote to eighteen of the world's senior naturalists, each with international experience and expertise in biodiversity and ecology, and asked their opinion on the best reserves, those sheltering assemblages of notably unique and valuable species of plants, animals, and microorganisms. I made the following request, here slightly paraphrased:

It is important to describe the real world and biodiversity in the way naturalists have experienced and known it, and which refutes that envisioned by de-extinction enthusiasts, nature-is-dead defeatists, and sundry ideological anthropocenists. Global biodiversity conservation should be judged and led by those who know it best. Not just that, but we need to ramp up the effort dramatically.

So here's the request: name one to five of the places in the world you consider best on the basis of richness, uniqueness, and most in need of research and protection, in other words those you most care about. And as you wish, give me reasons for your choice.

The "best places in the biosphere" are strongly personal and subjective choices. They are not the same as the global biodiversity "hot spots," as first defined in the 1980s by the British ecologist Norman Myers and others—although there is very considerable overlap. Hot spots are selected on the basis of the high number of species at greatest risk that would be saved by protection of the area in which they live. My consultants and I are well aware that our "best places" list could be multiplied many times over, but I believe that we have captured most of the very best. Together we make the point that even though extinction rates are soaring, a great deal of Earth's biodiversity can still be saved.

The Best Places Chosen

North America

THE REDWOOD FORESTS OF CALIFORNIA. Mark Moffett, an author and biodiversity expert, reports, "The most striking ecosystem within what's called the California Floristic Province is a hot spot for species diversity. I have climbed redwoods and sequoias with the researchers working in the region, and have been awed by the crowns of these trees, which are so ecologically lavish that groves of smaller trees, hidden from the ground, can grow out of the soils that accumulate on their immense branches." In effect, mature redwood forests have created a new and mostly unexplored layer of life, within which exist species rare or absent elsewhere. Scientists and adventurers can camp there, enthralled by the archetype of the giant tree that rises to a mythic world in the sky.

THE LONGLEAF PINE SAVANNA OF THE AMERICAN SOUTH. An increasing number of scientists and writers are turning their attention to this outwardly ordinary but remarkably rich and complex ecosystem. Once the dominant tree of 60 percent of the land from the Carolinas to eastern Texas, the longleaf savanna vegetation is adapted to frequent lightning-struck ground fires. The ground flora is one of the richest in North America, with as many as fifty herbaceous and shrubby species in a single hectare. Pitcher-plant bogs scattered through the longleaf woodland are in turn among the richest in the world, comprising up to fifty slender-stemmed species crowded into a square meter. The pine was almost entirely cut over during the past 150 years, but is now being reinstated to shelter the ground flora and fauna that survived its temporary disappearance. My own formative boyhood years were spent wandering through its remnants and the hardwood floodplain forests that dissect the region's countless rivers and streams.

THE MADREAN PINE-OAK WOODLANDS. The rugged Madrean mountain chains of Mexico and the Sky Island heights of the southwestern United States are dominated by low, dry forests of pine and oak. One-fourth of Mexico's native species occur in these ancient woodlands; many are found nowhere else. Pine forests in Michoacán are the famous wintering place of monarch butterflies from the United States. Of greatest importance is the role of woodlands as a corridor permitting north-south expansion of species between the United States, the Mexican Plateau, and the Cordilleras of Central America. Creating these and similar habitat corridors are one way of lessening the impact of climate change on biodiversity.

The West Indies

CUBA AND HISPANIOLA. The two largest islands of the Greater Antilles contain an exuberant fauna and flora that form the bulk of the biodiversity of all the West Indies. Their ultimate affinity is to Central America, from which the Antillean landmass broke away by continental drift tens of millions of years ago. Their long isolation has resulted in the origin of large numbers of species that today are found nowhere else. Some, like the strange insectivore mammal *Solenodon*, are relicts from the islands' earliest history. Others are the products of adaptive radiation, wherein one or a very few colonizing species have found less competition and more open niches than on the mainland. This circumstance has allowed some to proliferate into swarms of species that today individually fill the niches. Among the conspicuous examples are the diverse, abundant forms of anole lizards, and strange ants that glitter in metallic hues variously of blue and green. One day, while exploring in the Escambray Mountains of central Cuba, I found both a metallescent green ant species that nests in rock crevices and another species whose bodies flash gold foraging on low bushes. Moreover, both Cuba and the nearby Dominican Republic are surrounded by relatively undisturbed shorelines, with intact coral reefs.

South and Central America

THE AMAZON RIVER BASIN. This endless world of ecosystems, the largest drainage system in the world, also contains the greatest area of rain forest and the most biodiverse surrounding savannas. It feeds fifteen thousand primary and secondary tributaries of the

Amazon River and covers seven and a half million square kilometers, 40 percent of the area of the continent. Its biodiversity, if its Andean headwaters are included, is the largest in the world. The main stream commences in alpine rivulets of the Peruvian Andes. Running at an average velocity of 2.4 kilometers an hour and average depth of more than 45.7 meters, it moves thirty trillion liters of water each day through a deltaic mouth 402 kilometers wide. Its rate of discharge is eleven times that of the Mississippi River and sixty times that of the Nile. Its dimensions, when all of the tributaries are added, are matched by an immense diversity of fish and other freshwater animals and plants. The floodplain forests that clothe its banks, their ground and lower tree trunks underwater during the wet season, when combined with the inland terra firma rain forests, are reservoirs of an even greater diversity of animals and plants.

THE GUIANA SHIELD. The small countries of Guyana and Suriname, with neighboring French Guiana, are still 70 to 90 percent covered by pristine Amazon-related but distinctive rain forests. Their faunas and floras are exceptionally rich and remain among the least explored in the world.

THE TEPUIS. These tabletop mountains are the imagined "lost worlds" of H. G. Wells and Hollywood. Limited to Venezuela and western Guyana, they are composed of blocks of ancient quartz arenite sandstone thrust skyward from the surrounding rain forest. Their more or less flat tops range in elevation from one thousand to three thousand meters. They are indeed worlds unto themselves, with weather radically different from that at lower elevations, spectacular rock formations and waterfalls (Angel Falls is the highest in the world), and floras and faunas different from the lowlands below and from one tepui to the next.

GREATER MANÚ REGION OF PERU. Adrian Forsyth, a leading tropical biologist, has captured the magic of this region, "where the largest equatorial ice mass in the world, crowned by mighty Ausangate, towers above the rocky slopes and puna grasslands that turn into unroaded rain forests with the landscape then ranging down through trackless montane forest. It makes a compressed panorama visible in a single glance by anyone standing in the Amazonian lowlands of Madre de Dios." North of the river exists the most concentrated biodiversity on Earth, including a full array of New World larger mammals. A simple square kilometer can harbor the same number of species of frogs found in the entire continental United States, and twice the number of birds and butterflies. The same distinction or better applies to the rain forests of Ecuador's legendary Yasuni National Park, located still farther to the north.

CLOUD AND SUMMIT FORESTS OF CENTRAL AMERICA AND THE NORTHERN ANDES. These cool, rain-soaked environments are distinct in both climate and biodiversity from the lowland forests below. Many are poorly explored, and contain large numbers of still-unknown species. The finding of the olinguito, the first sizable carnivorous mammal discovered in a century, is emblematic of what remains hidden.

PÁRAMOS. These high-elevation (2,800 to 4,700 meters) grasslands of South America contain many unique grasses, herbs, and woody bushes. They also stand out in their high rate of evolution of new species—possibly due to fluctuations of climate in the fragmented mountaintop environment. Even though each is only kilometers above the lowland rain forests that surround them, they are physically and biologically a different world. Their floras are unique, and their small areas make their biodiversity vulnerable.

ATLANTIC FORESTS OF SOUTH AMERICA. Called the Mata Atlântica in Portuguese, this once vast and magnificent but now significantly reduced ecological system extends along the Brazilian Atlantic coast from Rio Grande do Norte state in the north to Rio Grande do Sul state in the south, with small extensions into Paraguay and Argentina (in other words, from the tip of Brazil's "nose" to southeastern Paraguay). Its latitude and local wide variation in rainfall give it an exceptional variation in ecosystems, which range from tropical to subtropical forests, both moist and dry, as well as scrub forests and grasslands. Within their domains occur a wide variety of rare and unusual animals, including, as described by Mark Moffett, "the most primitive porcupine; dancing frog and fruit-eating frog; the Alagoas curassow, the few remaining specimens of which roam the private properties of two bird lovers; the largest New World primate, the muriqui; the most colorful of all primates, the golden lion tamarin; the golden lancehead viper of Queimada Grande Island, where it reaches the highest density known for any snake (not surprisingly, no person lives on 'snake island')."

THE CERRADO. Covering a large part of east-central Brazil, the Cerrado is the largest savanna in South America and the most biodiversity-rich of any such tropical habitat in the world. Its luxurious variety of life derives from the sharply defined mosaic of ecosystems it comprises, from classic open grassland with scattered woodland copses to patches of tall, rain-forest-like trees, along with thick gallery forests that line its rivers. Unfortunately for biodiversity, its soil favors agriculture, and the Cerrado habitat is being rapidly cleared with as yet very little protection in reserves.

THE PANTANAL. One of Earth's largest wetlands, mostly in southern Brazil but extending into Bolivia, this magnificent floodplain is 80 percent underwater in the rainy season and is the year-round home to an immense variety of waterbird and insect life, as well as jaguars, capybaras, and other charismatic large mammals, including especially abundant crocodile-like caimans. Although designated as a World Heritage Site and increasingly popular with tourists, the Pantanal still supports a heavy burden of agriculture and cattle ranching.

THE GALÁPAGOS. This equatorial archipelago 926 kilometers west of the Ecuadorian mainland has acquired iconic status because of Charles Darwin's five-week visit there in 1835. Homeward bound, he noticed that mockingbirds differ from one island to the next, causing him to conceive of evolution. But the Galápagos are special for another reason: from just a very few species able to cross the ocean and colonize the islands have evolved many species that are specially adapted to the barren volcanic landscape. Giant tortoises, marine iguanas, herbs of the sunflower family that transmuted into trees, finches evolved from a single ancestor to fill a half-dozen bird niches, and more, make the islands both a laboratory and a classroom of evolutionary biology.

Europe

BIAŁOWIEŻA FOREST, POLAND AND BELARUS. This is the largest remaining fragment of the primeval forest that once covered the flatlands of northwestern Europe up to the dawn of the Neolithic period. Straddling the border between Poland and Belarus, the protected land covers almost two thousand square kilometers. A large fraction of the large mammals of Europe live within its confines, including most notably the European bison (which more than once has narrowly escaped extinction), roe deer, elk, wild boar, tarpan (a Polish wild forest horse), lynx, wolf, otter, and ermine. The nine hundred vascular plant species present include some of the largest oaks ever recorded.

LAKE BAIKAL, RUSSIAN SIBERIA. With an area of 31,722 square kilometers and maximum depth of 1,642 meters, Lake Baikal is the oldest and deepest freshwater lake in the world. As expected from its volume, it also harbors a remarkably large fauna and flora for an isolated body in the high latitudes. Of its approximately twenty-five hundred species of plants and animals, two-thirds are found nowhere else. Certain groups are represented by large numbers of species, including sculpins (fishes of the family Cottidae), sponges, snails, and amphipod crustaceans. Like the Galápagos Islands, Lake Baikal is both a sanctuary of biodiversity, in this case the most found in any freshwater lake, and a laboratory of evolution.

Africa and Madagascar

THE CHRISTIAN ORTHODOX CHURCH FORESTS, ETHIO-PIA. Less than 5 percent of the native forests of northern Ethiopia remain, and they are virtually all limited to properties of the Church—conspicuous from the air as green patches sprinkled across the brown landscape of subsistence agriculture. They contain, as Margaret Lowman has written, "the seed bank of native plants, pollinators for many garden crops, freshwater springs, medicines from plants, paints from fruits and seeds for the church murals, water conservation from roots, spiritual sanctuary as the church centerpiece, carbon storage, and home to the last remaining genetic library of native species."

SOCOTRA. An isolated island (with small satellites) located in the Indian Ocean 352 kilometers south of Yemen, Socotra is populated by trees and shrubs so strange in shape and foliage as to earn it the titles "another Galápagos" and "most alien-looking place on Earth." Here you will find the dragon's blood tree, the wall-dwelling Socotran fig, toothed aloes, and other plant species hard to compare even in general form with vegetation elsewhere. Socotra also has about two hundred bird species, of which eight are unique to the archipelago.

THE SERENGETI GRASSLAND ECOSYSTEM. Arguably the most famous terrestrial wildland ecosystem in the world is the great Serengeti (which means, in the Maasai language, "endless plains"), spanning a vast area from northern Tanzania to southwestern Kenya. A substantial portion of its area, as well as that of Kenya, is protected by national parks, conservation areas, and game reserves. The faunas and floras, including especially the large mammals, are the closest we have to the originals that have populated the African tropical grasslands and savannas since Pleistocene times.

GORONGOSA NATIONAL PARK, MOZAMBIQUE. The country's premier reserve has a full representation of southeastern African biodiversity due to the multifarious habitats found within it, including a mountain ridge rising to almost two thousand meters and topped by rain forest, miombo dry forest, multiple rivers and streams, and gorges with bottoms covered in rain forest and flanked by limestone cliffs, the latter riddled with still mostly unexplored caves. Gorongosa's megafauna is rapidly regenerating after being driven to near extinction during the civil war of 1978–1992 followed by widespread poaching.

SOUTH AFRICA. The country as a whole contains within its borders one of the world's several most abounding and distinctive assemblages of both animals and plants. Huge Kruger National Park in the northeast and other reserves are home to the most complete array of ten-kilogram-plus wildlife in Africa (including black and white rhinoceroses, both critically endangered). The Cape Floristic Region harbors nine thousand species, 69 percent found nowhere else, and one-fifth of all the plant species of Africa. Within the region are floras that form several unique major habitats, including the fynbos heathland, the succulent Karoo desert (shared with Namibia), and the archaic cycad forest of Limpopo Province.

FORESTS OF THE CONGO BASIN. The basin of the Congo River, its 3.4 million square kilometers spanning the Republic of the Congo, the Democratic Republic of the Congo, the Central African Republic, and parts of Cameroon, Gabon, Angola, Zambia, and Tanzania, is the second largest drainage system in the world, exceeded only by the Amazon. It is invested by tropical rain forest, constituting one of the world's three great rain forest wildernesses (the other two are the Amazon and New Guinea). Although under heavy siege from logging and conversion to agriculture, it remains home to more than three thousand unique plants and an immense fauna, including gorillas, okapi, forest elephants, and other famously spectacular species of large animals. Five of the rain forest parks of the Congo are United Nations World Heritage sites.

THE ATEWA FOREST, GHANA. Many of the moist forests of Africa's western hump have been reduced drastically by human encroachment, but fragments still survive as islands that preserve parts of the once exceptionally rich flora and forest. A splendid example is the pristine Atewa Forest, at least ten and a half million years old, a remnant of the rain forest that has elsewhere been 80 percent removed. Atewa is the finest example of a variant of the biome called the Upland Evergreen Forest.

MADAGASCAR. This huge island, the size of California and Arizona combined, located in the Indian Ocean four hundred kilometers off the coast of eastern Africa, has been isolated since it broke away 150 million years ago from the southern supercontinent of Gondwana. Being very big, ancient, and tropical, Madagascar harbors a very large and unique fauna and flora, with 70 percent or more of its species found nowhere else. (The most recent figure for the fourteen thousand plants is 90 percent.) Like Cuba, Hispaniola, and the Galápagos, Madagascar is a living laboratory for the observation of adaptive radiation, defined as the origin of a large array of species from a single species fortunate enough to colonize the island (in this case usually flying or floating over from Africa). Examples of evolutionary radiation in Madagascar's animals are the many, yet closely related, species of lemurs (primitive primates), chameleons, vangid shrikes, ranid frogs, and, among the twelve thousand species of plants, complexes of palms, orchids, baobabs, and cactus-like Didiereaceae.

Asia

THE ALTAI MOUNTAINS. Towering 4,509 meters at maximum elevation, this beautiful and seldom-visited mountain range rises in Central Asia where Russia, China, Mongolia, and Kazakhstan meet. Covered at different elevations variously by steppes, northern coniferous forests, and high alpine vegetation, it contains a living encyclopedia of cold temperate and arctic mammals, and is one of the few places in all of Eurasia that harbors a true Ice Age fauna. Herbivores abounding on its slopes include wapiti, moose, reindeer, Siberian musk deer, roe deer, and wild boar. Among their predators are brown bear, wolves, lynx, snow leopards, and wolverines. It is also the place where the first fossils of the Denisovan human species were found.

BORNEO. The 18,307 islands of Indonesia (the estimated number varies according to criteria and technology) straddle 5,120 kilometers of latitude, from the western tip of Sumatra to Irian Jaya, the western half of New Guinea. Together, the archipelagoes contain a staggering amount of biodiversity. The southern three-fourths of Borneo, the world's third-largest island, is part of Indonesia, while the northern one-fourth belongs respectively to Malaysia and the independent monarchy of Brunei. The island as a whole has lost a substantial fraction of its rain forest due to settlement and conversion to oil palm plantations. The damage has been what *Science* reported in 2007: "Palm oil plantations are spreading as sales of the biofuel soar, invasive acacia trees are on a rampage, and wildfires ravage the island each year." Yet the interior of the great island, the "Heart of Borneo," remains the leading single harbor of Asia's tropical biodiversity.

THE WESTERN GHATS OF INDIA. The subcontinent of India, like the great islands of Madagascar, New Caledonia, and New Zealand, is a fragment of Gondwana, differing primarily only in having drifted north all the way to meet and fuse with the mainland of Asia. The Western Ghats, a mountain range running parallel to the full length of the west coast, are the geological spine of India. Their range of elevation, from near sea level to 2,695 meters at the highest point, and their tropical location combine to create a great variety of terrestrial habitats, and with them a high level of biodiversity. Native to the mostly rolling, forested hills are five thousand species of plants, of which seventeen hundred are endemic, and a major mammal fauna that includes the world's largest population of wild Asian elephants and a tenth of Earth's surviving tigers.

BHUTAN. This idyllic mountain nation deserves praise for the preservation of much of its native habitats and biodiversity. Its faunas and floras remain an intact slice of those that originally characterized the bulk of the Himalayan mountains and foothills. In Bhutan, 70 percent of the land is covered by forest representing the three principal zones—tropical, temperate, and alpine. The five thousand known species of plants include forty-six rhododendrons and six hundred orchids.

MYANMAR. The northern reaches of this still-seldom-visited country harbor four reserves, about thirty-one thousand square kilometers in combined area, and a rich fauna that includes elephants, bears, red pandas, tigers, and gibbons. The region includes tropical forests, coniferous woodland, and even patches of arctic grasslands above tree line.

Australia and Melanesia

SCRUBLAND OF SOUTHWESTERN AUSTRALIA. From Esperance on the southwestern coast east to the edge of the Nullarbor Plain lies one of Earth's richest endemic floras. Possessing a mild, Mediterranean-type climate and molybdenum-deficient soil that excludes species other than those adapted to the deficiency, the scrublands have evolved much like the flora of an oceanic island. Unfortunately for the biodiversity of Australia and that of the world generally, molybdenum added to the soil makes it arable, and a large fraction of the scrubland has been converted to farms and cattle ranches, accompanied by incursions of invasive weeds.

THE KIMBERLEY REGION OF NORTHWESTERN AUSTRALIA. The national parks and other, remote areas of this part of the continent are among its most biologically diverse and least disturbed regions. Experiencing a comeback in its unique but elsewhere endangered marsupial fauna, Kimberley is correctly called Australia's "last great wilderness."

THE GIBBER PLAINS. The flat overflow lands of Sturt Stony Desert, located within the ultra-dry center of the continent, hold water only in the form of floods that occur many years apart. Abounding life emerges from dormancy at that time, drawing large flocks of waterbirds from far away. In between the errant rains the land is baked bone-dry, with knee-high sclerophyllous vegetation. Even when severe, it nevertheless teems with mostly hidden animal species. Bruce Means has written of the Gibber habitats, "More often than we realize, the miracle of biodiversity is right under our noses, wherever we are."

NEW GUINEA. The world's second largest island (after Greenland), about eight hundred thousand square kilometers in area, is still mostly invested in rain forest, wetlands, and upland grasslands. Its terrestrial biodiversity is generally recognized as both the richest and least explored in the world. The hyperdiversity is enhanced by the complex array of mountain ranges, the tallest soaring to alpine zones above forty-seven hundred meters, where occur permanently ice-covered peaks. New Guinea was an archipelago of smaller islands up to five million years ago, adding a secondary force that favored species formation. In 1955, as a twenty-five-year-old, I was the first to collect ants in a systematic fashion across parts of New Guinea. Today, if I were offered another six decades of healthy life but told I must spend it in one place as a naturalist, I would take New Guinea. No contest.

NEW CALEDONIA. This remarkable island, subtropical and mountainous, has been isolated as a Gondwanan fragment for eighty million years. It was first joined with New Zealand, then broke off to drift completely alone toward the equator. Today more than 80 percent of the native plant and animal species are found nowhere else, and many are markedly different from anything found elsewhere. New Caledonia even harbors elements that were present before it departed the Australia-plus-Antarctica landmass. It has the highest number of endemic families of plants, including those with archaic features, most famously *Amborella*, the most primitive flowering plant known on Earth. Along mountain ridges are still mingled forests of *Araucaria* and *Podocarpus*, providing an environment resembling that which prevailed over most of Earth during the Mesozoic Era. I conducted research on the island during 1954 when I was a graduate student, then often had dreams of it during sleep, until I returned fifty-seven years later, in 2011. The magic was still the same for me in real life that it had been in 1954.

Antarctica

MCMURDO DRY VALLEYS. Here, in the most inhospitable ice-free land on Earth, a place rivaled in the poverty of its biodiversity only by the sere, rainless Atacama Desert of Chile, live just enough species to make a balanced ecosystem. Sparse traces of algae are the plants, and several species of nematodes, also commonly called roundworms, are variously herbivores and predators—what Diana Wall at Colorado State University calls the "elephants and tigers" of the Antarctic soil. The simplicity of the material and energy cycle reminds us that there exist organisms able to establish themselves almost anywhere. But as humanity scales down Earth's ecosystems, life will become progressively less interesting and more difficult to turn into a support system.

Polynesia

HAWAII. The Hawaiian archipelago, like the equally far-flung Easter, Pitcairn, and Marquesas archipelagoes, deserves mention in part for what it once was. Its tropical climate, relatively large size, and mountainous terrain with multitudinous habitats promoted the genesis of a large diversity of land-dwelling plants and animals. A high percentage of these originated as products of adaptive radiations. Dramatic examples of such species swarms include the honeycreepers among the smaller birds, tree crickets among the insects, and lobelias among the flowering plants. The beautiful assemblage has been largely wiped out or pushed into the remote uplands of the central mountains by agricultural conversion and semiwild gardens of invasive species. Perversely, the latter have become a poster child for the "novel ecosystems" celebrated by Anthropocene supporters.

There is nevertheless a "best place" on the Hawaiian archipelago. It has been movingly described by Stuart Pimm of Duke University, a leading expert on extinction and remnant native birds:

> Looking out from just above the tree line on Maui, the forest seen immediately below is short, stunted, very wet, and only on very rare clear days can one see the lowlands filled with tourists—and almost entirely alien trees. But here you can imagine a world that was like no other—all its species were endemic. This is a separate evolution—birds with strange names, 'akohekoe, o'o, 'akialoa, nuku pu'u—and even stranger beaks and endemic lobelias into which they stuck them—and yes, no ants. It's a forest of ghosts—only the 'akohekoe survives—but what remains is a precious reminder of how special it was, what we

must do to keep what remains, and what we must not allow to happen elsewhere.

In the mind's eye it is possible to link the premier terrestrial wildlands into a single global circuit, and envision traveling in a near-continuous journey through it all. Within this broken circle of life the natural world can be seen as it was ten thousand years ago, when humanity still occupied the planet thinly and only in small parts, and agriculture was new and sparse.

Such a trip is the reverse of ordinary present-day travel. Instead of passing over wildlands to skip from city to city, it passes over cities to get to the wildlands.

The odyssey is more instructive if the pathway chosen is the same taken by our ancestors starting more than sixty thousand years ago. The trail begins in the birthplace of humanity, the savannas and miombo dry forests of southern and central Africa, much of which is still in wilderness condition. It takes a side trip to the rain forests of the Congo Basin and West Africa. Then north along the Nile, and perhaps also across the Bab-el-Mandeb, out of Africa and into Eurasia. By necessity the trail is broken by the densely settled lands of the broad Mediterranean region, including all of the Middle East. It takes up again the Białowieża Forest of Poland and Belarus, the largest primeval woodland remaining in the middle latitudes of Europe. Then it enters the taiga, the northern coniferous forest that begins in Scandinavia and Finland and stretches mostly unbroken eastward seven thousand kilometers across the Eurasian Supercontinent to the Pacific. Along the way it passes Lake Baikal, the largest body of freshwater in the world and home to the most endemic species of north temperate aquatic animals.

From the Amur River region of Siberia and northern China the wilderness trail jumps to the Altai Mountains of Central Asia, the Tibetan Plateau, and remote parts of the Himalayan southern face. It picks up again in the montane and tropical forests of Myanmar and the Western Ghats of India.

The trail continues through Indonesia and the rapidly diminishing number of its islands that remain undisturbed. Eastward it enters the densely forested island of New Guinea in both of its political divisions, Indonesian Irian in the west and the independent country of Papua New Guinea in the east. From the southernmost string of Indonesia's Lesser Sunda Islands, plus Timor-Leste, the trail then crosses the Timor Sea, as did the first aborigines to reach Australia's Northern Territory and the Kimberley Region in the southern continent's northwest. Both remain mostly in their original ecological condition.

The trail breaks and takes up again in far northeastern Siberia, crosses over by the still mostly empty Aleutian Islands to Alaska, passes inland to the immense span of arctic and subarctic scrub habitats, then southward into the Canadian taiga. In the western part of the continent, it passes down along the coast to reach all the way to the still well preserved mountain habitats and lowland tropical parts of Central and South America, thence inland to substantial blocks of original habitat. Finally, from the Peruvian highlands to Belém, through rain forest to tropical grassland, it runs along the mightiest river fed from the largest river basin in the world.

Our fragmented circle of life ends in the eastern slopes and foothills of the Andes, where we find both the final continental region to be reached by human beings and largest number of wild species of plants and animals that grace a single place.

Two waterleafs, Wigandia crispa *and* Hydrolea diatoma *of Peru.*
Hipólito Ruiz López and Josepho Pavon, 1798–1802.

16

HISTORY REDEFINED

History is not a prerogative of the human species. In the living world there are millions of histories. Each species is the inheritor of an ancient lineage. History exists in a point of space and time after a long journey through the labyrinth of evolution. Each twist and turn has been a gamble with the species' continued existence. The players are the many ensembles of genes in the population. The game is the navigation of the environment in which the population lives. The payout is the share of breeding individuals in the next generation. The traits prescribed by the genes that sufficed in past generations might in the future continue to do so, but might not. The environment is also changing. In new environments the genes may keep on winning, allowing the species to survive. Or not. Some of the variants of the genes, having arisen by mutation or forming new combinations, might even cause the species population to grow and spread. But at any time in a changing environment, the species could lose this game of evolution, and its population would spiral to extinction.

The average life span of a species varies according to taxonomic group. It is as long as tens of millions of years for ants and trees, and as short as half a million years for mammals. The average span across all groups combined appears to be (very roughly) a million years. By that time the species may have changed enough to be called a different species, or else it may have split into two or more species—or vanished entirely to join the more than 99 percent that have come and gone since the origin of life. Keep in mind that every surviving species (including us) is therefore a champion in a club of champions. We all are best of the best, descendants of species that have never turned wrong in the maze, never lost. Not yet.

The history of a species is thereby an epic. In time, maybe not in this century given the huge number of other species still alive in the biosphere, scientists might get around to studying the biology of any particular randomly selected species in total depth. They will explore its life cycle, its anatomy, physiology, genetics, and ecological niche. They will learn as much as possible about its geological history. Fossils, when available, will help them immensely. But more likely, its history will be inferred by comparisons with other species that most closely resemble it. The idea is to place it in the family tree of the most closely related species. With the aid of DNA sequencing, the researchers will determine with which living species it arose. Their common ancestors will be traced back, as in human genealogies, hence inward through many twigs and branches. The genetic analysis, joined to the evidence of the species' inherited biological traits, will reveal where the relatives now live, where they lived in the past, and what their biological traits may once have been. An evolutionary family tree of this kind is called a phylogeny. Its reconstruction may be the closest we will ever get to telling the species' epic. But as more and more such stories are written, principles in the history of life will be clarified. The living world around us will make more and more sense, continuing for all time to come.

The human species, of course, has an evolutionary history, which reaches very far back in time beyond traditional recorded history. We, too, are the twig-end of a phylogeny. The multitudinous stories of human cultures are epics in the usual sense, but you will understand that the traits of human nature that have molded these stories are also products of evolution. We have our australopith cousins and our australopith forebears, together with grandmother *Homo habilis* and our mother *Homo erectus*. The two levels, biological and cultural, flow one into the other. This is the reason that history makes no sense without prehistory, and prehistory makes no sense without biology.

Looking back to deepest geological history, to the primordial single-celled prokaryotes, all living species are seen to be members of the same phylogenetic megafamily. Tracing forward from 3.8 billion years ago to a point 55 million years ago, we find the limb (so to speak) of all the Old World primates. Pressing forward and outward, we come to the branch of the hominids, and finally to humanity.

Our key adaptation, our stroke of evolutionary good luck, is our relatively powerful minds. With it we re-create episodes of the past. We invent alternative episodes for the future, select one of them, and perhaps decide to make it part of our story. We are the only entity on the planet that accumulates knowledge for its own sake and, by combining it with the knowledge and cooperation among ourselves, makes decisions for the future, often wise but equally often disastrous.

And so it has come to pass that we have chosen to learn all we can about the rest of life—all of life, the whole biosphere. To discover every species of organism on Earth and to learn everything possible about it is of course one of the most daunting of all tasks. But we will do it, because humanity needs the information for many basic scientific and practical reasons, and more deeply and compellingly because exploration of the unknown is in our genes. In time the mapping of Earth's biodiversity will become a Big Science project, comparable to cancer research and the brain activity map prevailing at the present day. Unless our current estimates of biodiversity are wildly off, there are about a thousand people alive on Earth for every species of organism. In theory a sponsor might easily be found for each and every one. The collective human mind, hyperconnected and digitized, will flow through the entirety of the life we have inherited far more quickly than was possible before. We will then understand the full meaning of extinction, and we will come to regret deeply every species humanity will have carelessly thrown away.

All biological knowledge begins with names and classification. When specimens can be identified to a species, every bit of information accumulated about that species becomes immediately available. There is magic in the Linnaean double name, such as *Drosophila melanogaster* for the common fruit fly and *Haliaeetus leucocephalus* for the bald eagle. It is the key to finding everything that has been learned by science about the species. It conjures up what we personally know or think we know. The double name forms the basis of a hierarchy suited to the way the human mind actually works. In repeating it over and over, listening to the sound of it and sensing the unknown, it is the poetry of science.

Biologists define a species as a collectivity of individuals that breed with one another under natural conditions. I've earlier cited as an example the lions, which bear the scientific name *Panthera leo*. The lions are close relatives of the tigers, *Panthera tigris*. The name *Panthera* denotes the genus comprising those species of big cats that are genetically close to one another and to lions and tigers. The classical system of taxonomic classification then continues upward and outward, in hierarchical order, as in leaf to twig to branch. The cat species of the genus *Panthera* when put together with other cat genera (genera is the plural of genus), for example those that include the domestic cat, the wildcat, lynx, and the jaguarundi, form the taxonomic cat family, the Felidae. The species of the Felidae are joined with the dog family, the Canidae, and put with other related mammalian families to form the order Carnivora. And so on up the taxonomic hierarchy until all the species of animals, plants, and microorganisms, both living and extinct, have been included.

This is old-fashioned taxonomy, dating back more than 250 years to Carl Linnaeus, and it works. It provides the foundation and framework of classification, as well as the language of scientific natural history. The two-part scientific name itself is based semantically on Latin and Greek, and is spoken across all cultures. It provides each specimen with a name as we do for a person, and it leads easily to all the levels in the hierarchy to which the specimen belongs. All knowledge accumulated to the present moment about and around the species is made available by citing the name of the specimen.

The taxonomic name flows with the way the brain works and the way we all most readily communicate. We can talk easily using the Linnaean double names. A field biologist might say of a tiny fly circling over a banana the following: "It's a fruit fly, a drosophilid, almost certainly a member of the genus *Drosophila*. My guess is that it's *Drosophila melanogaster*, but to be completely sure I'd need to check its key traits under a microscope." And of a small spider with splayed-out legs and two elongated spinnerets extended like a tail behind it, resting on the bark of a red mangrove tree (*Rhizophora mangle*): "This is a hersiliid. I have no idea yet what genus or species of the family Hersiliidae it belongs to, but that's unmistakably a hersiliid." And next, an elongated creature with multiple legs. "That is a centipede. And not just any centipede, but a member of the family Lithobiidae. It's a lithobiid and definitely not a scolopendrid, or a scutigerid, or a geophilid, or a species of any of the ten other known families of centipedes found around the world."

Finally, in Mozambique's Gorongosa National Park, where I have conducted recent field research, we see a solid column of large ants beginning to march across a dirt road. I might explain then to a visitor:

> Those are matabele ants, as they're called in this part of Africa, after the Matabele warriors of old Zimbabwe, and we're beginning to study them here more carefully. *Pachycondyla analis* is the correct Latinized name, and it's the only ant known that marches in a highly coordinated column all running in the same direction like this one. It has to march like that if it's going to get any food. It preys exclusively on termites, and like every species of termite has a strong soldier caste guarding the nest entrance. The ants easily defeat the termite soldiers, and each one gathers up to several of the termites in its jaws to carry back to its own nest. Amazing! The matabele ants go

to battle for one reason only. Their diet consists exclusively of dead termite soldiers.

The knowledge accumulating in scientific studies of any species is organized by its place in the hierarchy. Its genetic relationships and evolutionary history are given by its place, and its name can be changed when new evidence forces a change in that place. If this hierarchical system did not exist, and was not subjected as it has been to the strict, internationally sanctioned rules of zoological and botanical literature, knowledge of Earth's biodiversity would quickly descend into chaos.

The hierarchical system and formal nomenclatural rules cannot be easily changed, but the transmission of information using it has been vastly improved by the digital revolution. For most of my career, in which taxonomic studies of ants played a major part, I had no choice but to borrow reference specimens on which the species names and classifications are based, or else visit the far-flung collections in museums throughout Europe and America in which they are deposited. To consult the literature, I had to search further through antiquated or highly specialized journals. I was fortunate to be at Harvard University, with the largest collection of ants, with perhaps seven thousand species and millions of preserved specimens—too many to count! It also has one of the best zoological libraries in the world. I had to travel less than my colleagues. But taxonomic research was still a slow process.

The crippling bottleneck I've just described has now been largely eliminated throughout the classification of all animals, algae, fungi, and plants. Crucial specimens, including the "type" specimens on which the name was originally based, are photographed with a high degree of resolution. Their three-dimensional traits are made clear by computer software. The images are then uploaded to a website with a description and citations so that anyone else in the world can see them with a few keystrokes. The entire literature of biodiversity is currently being scanned and will be made available online by a consortium of major universities and research institutes. The final product, called the Biodiversity Heritage Library, will eventually contain as many as five hundred million pages. Meanwhile, the Encyclopedia of Life, designed to summarize and present free most available information on all described species from its website, is well along, with the number of pages at the time of writing (2015) approaching 1.4 million, more than 50 percent of all the known species in the world. Complementary projects with additional knowledge, which grow as data are added, include the Global Biodiversity Information Facility, Map of Life, Vital Signs, USA National Phenology Network, AntWiki, FishBase, and not least GenBank, the immense, publicly accessible repository of DNA sequences. In a nutshell, the digital revolution has propelled the classification of life forward by decades, perhaps centuries.

As the databases in all these enterprises grow, new methods are being developed to convert their content into search engines to aid the rapid identification of specimens. By far the most potent is barcoding. The key is the DNA sequence of mitochondrial genes, which lie outside the nucleus in each cell, hence are inherited only through the mother. One segment in the gene COI, comprising only six hundred fifty base pairs, is particularly useful because it varies consistently from one species to the next. Read COI, and in most cases, perhaps all cases, biologists can name the species—providing it is known to science. With this method they can also match very dissimilar life stages, such as caterpillars with the adult butterflies into which they metamorphose. In the manner of forensic science, they can even identify tiny fragments of organisms to the correct species. And for the first time, it is possible to distinguish species so similar in anatomy that they cannot be separated by standard taxonomic methods.

Good things, however, invite overenthusiasm, and such is the case for barcoding. Some of its users see it as the solution to a shortage of expert taxonomists in the scientific world, as well as a direct route to the mapping of global biodiversity, and even a replacement of the prevailing hierarchical, name-based system of classification. However, these hopes are conspicuously in vain. The barcoding method is a technology, but it is not an advance in science or scientific knowledge.

Furthermore, there is no guarantee that the exploration of Earth's biodiversity can be completed before the twenty-third century. The problem is a severe shortage of expert researchers. Technology without science is like an automobile without wheels and a road map. The solution to the problem is more naturalists, or more precisely, scientific natural historians. We need many more experts on particular groups of organisms, dedicated specialists who conduct original research on the classification of the species, their natural history, and, in collaboration with other scientists, ultimately all of the biology of the species in their favored group. These scientists are furthermore historians, custodians of the stories each species will tell as its biology unfolds. Scientific naturalists were once among the leaders of biology. They were, and still remain in smaller numbers, masters of the *logos* ("reasoned discourse": Aristotle), mammalogists on mammals, herpetologists on reptiles and amphibians, botanists on angiosperms, mycologists on fungi, and so on through the long Linnaean roster. Their numbers have been cut in the erroneous belief that the living environment is less important to humanity than the nonliving environment.

The scientific naturalists were and remain a special breed. They do not select a special process or outstanding problem on which to focus their attention. They are not prepared to devote their careers to tracing a biochemical cycle, penetrating strata of a nuclear membrane, mapping brain circuitry, or achieving some other comparable major goal. Instead they are out to learn everything, by which I mean literally *everything*, about the biology of the group they have chosen. The group may not be all the birds, but it could be the passerine birds of South America; maybe not all the flowering plants, but, say, the oak species of eastern North America. Every scrap of information is deemed valuable and publishable, somewhere, even if just online.

Naturalists are thereby the ones who make the most surprising discoveries and often also the most important ones. They routinely encounter phenomena that other biologists, absorbed in the minutiae of molecular and cell organization of a few dozen model species, never even imagine. I admit I felt a measure of pride in my own work and that of my colleagues when on one occasion I was introduced as the keynote speaker of a major conference on behavioral biology by an illustrious molecular biologist, who said, "Ed is the kind of biologist who makes the discoveries on which we work."

An authentic scientific naturalist is devoted to his group of species. He feels responsible for them. He *loves* them, not the literal earthworms or liver flukes or cavernicolous mosses he might be studying, but the research that reveals their secrets and the place of these chosen organisms in the world. It has long been clear to me that biologists are divided into two kinds of people, differing by worldview and research methodology. The first tribe follows the rule that for every problem in biology there exists an organism ideal for its solution. And thus are selected the model species: fruit flies for particulate heredity, the gut bacterium *Escherichia coli* for molecular genetics, the roundworm *Caenorhabditis elegans* for architecture of the nervous system, and so on across the long full arc of molecular biology, cellular biology, developmental biology, neurobiology, and of course biomedicine. In contrast, for the second tribe, the naturalists, the rule is the inverse of the rule for the first tribe: for every organism there exists a problem for the solution of which it is ideal. There have been stickleback fish for instinctive behavior, cone mollusks and poison-arrow frogs for neurotoxins, ants and moths for pheromones, and thence through every level of biological organization, from cell assembly through all the principles of organismic and evolutionary biology.

Unfortunately, there has been more competition than cooperation between members of the two tribes, and the naturalists have decisively lost. Since the 1950s, when molecular biology was born and the golden age of modern biology was inaugurated, funding and prestige have shifted massively to a structural-biology and model-species school. A large part of this support has come from the obvious relevance to medicine. From 1962 through the rest of the century, the second tribe, organismic and evolutionary biology, saw its proportion of Ph.D.s drop precipitously, while that of microbiology, molecular biology, and developmental biology soared. The number of faculty positions in research universities followed the same trend, despite the obvious relevance of natural history and biodiversity studies to ecology and the remainder of environmental science. Scientific naturalists, often erroneously labeled as old-fashioned and as largely washed out, found some refuge in museums and environmental research organizations, but even these positions became less secure and well funded with time.

This disparity in the prestige and support of disciplines is a loser for science and for humanity's ability to protect the living environment. If ecology and conservation biology are ever able to mature enough to conserve Earth's biodiversity, it will be done not by theory and high-altitude overflights of the ecosystems, not by studies of molecular and cellular biology, but by taxonomic boots on the ground. Let great credit continue flowing to those who explore the broad traits of habitats and intricate details of a few kinds of bacteria, roundworms, and mice, but also to those dwindling few who study everything else.

PART III

The Solution

The global conservation movement has temporarily mitigated but hardly

stopped the ongoing extinction of species. The rate of loss is instead

accelerating. If biodiversity is to be returned to the baseline level of

extinction that existed before the spread of humanity, and thus saved

for future generations, the conservation effort must be raised to a new

level. The only solution to the "Sixth Extinction" is to increase the area

of inviolable natural reserves to half the surface of the Earth or greater.

This expansion is favored by unplanned consequences of ongoing human

population growth and movement and evolution of the economy now

driven by the digital revolution. But it also requires a fundamental shift

in moral reasoning concerning our relation to the living environment.

The velvetfish Aploactis milesii *(above) and goblinfish* Glyptauchen panduratus *(below).* Proceedings of the Zoological Society of London, *1848–1860.*

17

THE AWAKENING

Earth is a Goldilocks planet, not close enough to its star to be burned to a crisp, not far enough away to be locked in eternal ice. Life began on this planet more than three billion years ago, yet on Antarctica there is part of the biosphere that could not advance. Elsewhere, and even close by in the icy waters of the Antarctic shallow sea, life flourished and diversified. But inland, at Lake Untersee in the mountains of Queen Maud Land, possessing an environment closer to that of Mars than of Earth, evolution stalled. Dale Andersen, a research scientist at the SETI Institute, has described Lake Untersee as "a place few people have seen or even imagined."

The weather can be as harsh as the terrain at times with winds and blinding snow reaching 110 mph. For four months the darkness is company only to the sounds of cracking ice and the ever present howling of the winds. The surrounding mountains rise up majestically to soaring, pinnacled peaks, blocking the passage of the continental ice that surrounds them. The gentle slope of the Anuchin Glacier flows from the north, halting at its edge. Lake Untersee within the mountains of Queen Maud Land is a world that resembles Earth's earliest biosphere. One dominated by microbial life forming the same fabrics and structures that we see preserved in sediments dating back 3.45 billion years. Beneath the thick perennial ice-cover are cyanobacterial mats growing undisturbed as they did billions of years ago . . .

Now suppose that the Antarctic terrain had been by happenstance the best for life with a different code and a different mode of star energy and mineral energy capture. The norm and the efflorescence of biodiversity might then have been as it is in Queen Maud Land, and the equatorial belt of the Amazon and Congo too hot to sustain anything but marginal, primitive life.

It helps perspective to look at Earth as a whole this way. If the approximately one billion years of evolution it took for the single-celled bacteria and archaea on our planet to evolve into more complex life-forms were added, it is possible to sense how delicate our birthplace is, how complicated those parts of the ecosystems that shelter each species are, and how intricate and intertwined are the nonlinear interactions of the species. Earth's biosphere is like the orb of a spiderweb through which a bird has accidentally flown. Beautiful order has instantly turned into chaos. The spider knows this risk instinctively, and builds across part of the web a band of silk so conspicuous that it virtually shouts at intruders to turn aside.

Warning signs like that of the spider are all around us, yet the Darwinian propensity in our brain's machinery to favor short-term decisions over long-range planning makes us ignore them. I'm reminded of a conversation I had in 2005 with a hydrologist at Texas Tech University. I had been impressed by the rich agriculture of the Texas Panhandle but aware of its dependence on irrigation with water from the Ogallala aquifer. Knowing that the water recharges at a much slower rate than its current withdrawal, I asked my companion how long it would last. "Oh, about twenty years, if we're careful." I said, "What will you do then?" He replied, with a shrug, "Oh, we'll think of something."

I hope he's right, but the signs are not favorable. Here and elsewhere in marginal habitats, climate change and myopic thinking have taken a heavy toll. As deserts creep relentlessly across the Sahel of Africa, as Australia's arid center presses outward against the coastal croplands, and as the Colorado River has no more to give to the water-starved fields of America's Southwest, agriculturists will at last have to turn to dryland crops. They will need species characterized by deep perennial roots, and more drought-resistant grasses that bear edible seeds.

The world as a whole is already well into a water crisis. About eighteen countries, home to half the world's population, are draining their aquifers. In Hebei Province, in the heart of China's northern grain belt, the average water level in the deep aquifer is dropping nearly three meters a year. Underground water levels are falling so fast in the lowlands of rural India that in some localities drinking water must be trucked in. One official of the International Water Management Institute has said, "When the balloon bursts, untold anarchy will be the lot of rural India." In the Middle East, it is becoming clear that hatred and instability are not due so much to religious differences and the memories of historical injustice as they are to overpopulation and the severe shortage of arable land and water.

Earth's more than seven billion people are collectively ravenous consumers of all of the planet's inadequate bounty. Ten billion, give or take a billion, expected by end of the century, will be even more ravenous unless agrobiology and high technology can somehow turn the tide. Agriculture faces other realities. At present we consume nearly one-quarter of Earth's natural photosynthetic productivity: that much of the planet's freshly manufactured biomass ends up in our hands and stomachs and the share is growing. The remainder of the planet's productivity is left for all of the other millions of species.

The overall productivity of Earth at the present time can be summarized precisely as follows. For the past thirty years at least, as Steven W. Running of the University of Montana has reported, the land-based and primarily plant-based net primary productivity (NPP) of the planet has remained almost constant, varying by less than 2 percent annually. Total global precipitation has varied by only 2 percent, and the global input of solar radiation, which drives photosynthesis, has fluctuated by less that 0.01 percent. Humans now co-opt for energy and fuel about 38 percent of the NPP. Can humans continue to increase and consume in a way that allows us to take over the remaining 62 percent? No, I'm afraid not, at least not through conventional agriculture. If the unharvestable part is subtracted, it leaves only 10 percent of total global NPP available for additional use by humans, and that fraction is located mostly in Africa and South America. Unless a new Green Revolution can be engineered, human use will risk elimination of most of the remaining land-based biodiversity.

The pivotal conclusion to be drawn remains forever the same: by destroying most of the biosphere with archaic short-term methods, we are setting ourselves up for a self-inflicted disaster. Across eons the diversity of species has created ecosystems that provide a maximum level of stability. Climate changes and uncontrollable catastrophes from earthquakes, volcanic eruptions, and asteroid strikes have thrown nature off balance, but in relatively short geologic periods of time, the damage was repaired—due to the great variety and resilience of the life-forms on Earth.

Finally, during the Anthropocene, Earth's shield of biodiversity is being shattered and the pieces are being thrown away. In its place is being inserted only the promise that all can be solved by human ingenuity. Some hope we can take over the controls, monitor the sensors, and push the right buttons to run Earth the way we choose. In response, all the rest of us should be asking: Can the planet be run as a true spaceship by one intelligent species? Surely we would be foolish to take such a large and dangerous gamble. There is nothing our scientists and political leaders can do to replace the still-unimaginable complex of niches and the interactions of the millions of species that fill them. If we try, as we seem determined to do, and then even if we succeed to some extent, remember we won't be able to go back. The result will be irreversible. We have only one planet and we are allowed only one such experiment. Why make a world-threatening and unnecessary gamble if a safe option is open?

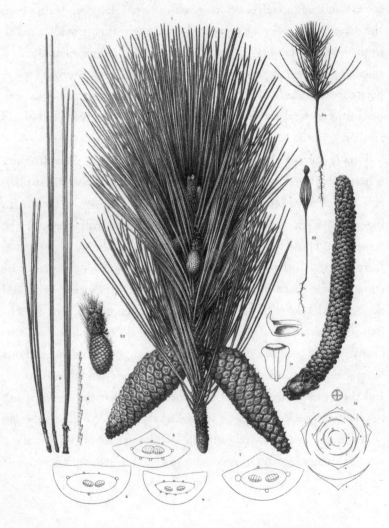

Slash pine (Pinus elliottii) *of the southeastern United States.*
George Engelmann, 1880.

18

RESTORATION

There are true wildernesses around the world that, if simply left alone, will endure as wildernesses. In addition there are mostly wild places whose living environments can be returned close to their original condition, either by the removal of a few invasive species or the reintroduction of one to several extirpated keystone species—or both. At the opposite extreme are landscapes so degraded that their original life must be restored from the ground up, by inserting soil, microorganisms, and eukaryotic species (algae, fungi, plants, animals) in certain combinations and in particular sequences.

For a large minority of conservation projects, some amount of restoration, meaning human intervention, is necessary. Each project is special unto itself. Each requires knowledge and love of the local environment shared by partnerships of scientists, activists, and political and economic leaders. To succeed, it needs every bit of their entrepreneurship, courage, and persistence.

Large conservation programs, like new scientific disciplines, start with a heroic age. A few individuals push forward, risking failure and harm to their own security and reputations. They have a dream that does not fit the norm. They accept long hours, personal expense, nagging uncertainty, and rejection. When they succeed, their idiosyncratic views become the new normal. Their individual stories are then rightfully seen as epics. They become part of environmental history.

During my work in nature parks and reserves, I have had the privilege of working with two such pioneers in biodiversity conservation. Their heroic endeavors unfolded on different continents. The landscapes they restored had problems that at first seemed polar opposites, but these conservationists were motivated by the same prime movers: the love of the habitats they befriended and the perceived need to bring back keystone species earlier destroyed by human action.

MC Davis of Miramar Beach, Florida, was (until his death from cancer in 2015) a very successful business entrepreneur whose wealth was built in good part from property management and the rehabilitation of small businesses. At first his way of life, focused on capital investment and development, seemed typical for an American businessman. But he was also an outdoorsman who explored the wild environments of his native Florida Panhandle with a special passion for science and education. Learning ecology and natural history on his own, MC discovered that the biodiversity of most of the Panhandle woodlands was in a seriously disturbed condition. The principal cause, he learned, was the disappearance of the longleaf pine (*Pinus palustris*), the signature tree of the southern U.S. wildlands.

The longleaf pine is a tall, stately tree that yields lumber of high quality, ranked with white pine and redwood among the best in the

United States. Before the coming of Europeans, it was the dominant species on 60 percent of the Southern wildlands. Longleaf was not tightly arrayed into a forest, nor was it the most abundant tree within small hardwood forest groves scattered across the landscape. Rather, it ruled over an open savanna. Other tree species in its midst were kept sparse by frequent lightning-sparked fires. Longleafs survived because they evolved special resistance at the seedling stage, including rapid aboveground growth and deep roots. It is relatively easy to walk through an old-growth stand of longleaf pine, because the understory consists primarily of low herbs and shrubs, representing a great variety of flowering species also adapted to survive the frequent fires.

Following the U.S. Civil War, Northern entrepreneurs and newly impoverished Southerners began harvesting longleaf pine as a principal source of income. By the end of the twentieth century, less than 1 percent of the pristine original cover remained.

The clear-cutting resulted in more than the severe reduction of the dominant species. It changed the entire structure of the remaining savanna. Previous "weedy" tree species, including fast-growing slash pine and loblolly pine, took over from the commercially more valuable longleaf. Higher understory shrubs replaced much of the original species-rich understory. They and the newly dominant pines allowed the accumulation of thick, dried-out leaf litter and masses of flammable dead branches caught and held well above ground level. The result was that natural fires no longer crept along close to the ground and burned themselves out among vegetation adapted to resist them. Aided by even a slight wind, they instead swept upward through the understory and raged outward through the canopy as full-blown wildfires. I know this degraded environment very well. I spent the bulk of my boyhood wandering through it in southern Alabama and the western Florida Panhandle. Only as an adult did I understand the full picture of its decline.

MC Davis recognized that restoring the longleaf pine was the key to the health and sustainability of the Florida Panhandle and a substantial region of the southern United States beyond. Other environmentalists, including forestry professionals in the Longleaf Alliance and similar environmental organizations, had also grasped the problem well and begun to address it. But it was Davis who as a private citizen chose single-handedly to do something dramatic about this problem. And he did. He noticed that undeveloped property away from the beach zone of the Gulf of Mexico coast, having been shorn of its longleaf and left with soil relatively poor for farming, could be purchased cheaply. In partnership with a fellow business entrepreneur, Sam Shine, he bought up large sections of land and put them into a permanent conservation trust.

Then came the even harder part, restoration of the longleaf savanna. Davis acquired big-timber equipment and set out to cut down the intrusive slash and loblolly pines. He arranged to sell the timber to help pay for the operation. His crews used other, specialized machines to claw out the dense fire-prone parts of the understory. When the cleared land was ready, they planted more than a million longleaf seedlings. Where the South's premier timber species began to return, the rich ground flora resumed its original unimpeded growth.

In the course of restoring to life a part of northern Florida's original habitat, MC Davis conceived another idea. While we're at it (he said in a good-old-boy Deep Southern drawl), we might create a wildlife corridor, consisting of a narrow but continuous strip of natural environment along the Gulf of Mexico coast from Tallahassee west to Mississippi. The corridor would allow larger animals, including bears and cougars, to reoccupy regions from which they had disappeared decades ago. It would also make possible a moderation of damage due to climate change, which was expected to include an eastbound wave of drying along the Gulf Coast. Such a linkage is now viewed as a possibility. Even better, its creation is under way. The parts, achieved and planned, include state and federal woodland, coastal river floodplain forests, military buffer property, and privately owned wildland tracts.

Gregory C. Carr, scion of a pioneer family in faraway Idaho, is the second American entrepreneur restoring wildlands whom I've had the privilege of knowing and assisting. Wealthy from innovation in telephone voice technology and its commercial development, Carr devoted his life to the gargantuan task of restoring the Gorongosa National Park in Mozambique. During the country's civil war of 1978 through 1992, which left as many as a million people dead, and continuing through the heavy poaching that followed, almost all of the megafauna, including elephants, lions, and fourteen species of antelopes, were extinguished or brought to the brink of extinction. On the once-sacred slopes of Mount Gorongosa, local people had begun to cut away the rain forest, the main rainfall catchment of water for the park and the surrounding area.

Following his first visit, on March 30, 2004, Greg Carr set out to restore Gorongosa to its original condition. He rebuilt the central camp at Chitengo and added, as a wholly new feature, a laboratory and museum for the thorough study of the fauna and flora of the park and surrounding area. By the end of the first decade he had largely fulfilled his original goals. And, as planned, tourists were returning in increasing numbers.

Carr's innovation was by no means narrowly focused on science and conservation. From the beginning, he also gave high priority to the welfare of people living in and around Gorongosa. Hundreds of local people were employed in the park, from laborers and construction workers to restaurant workers and rangers. A Mozambican was named park warden, and in addition functioned as the official liaison with the central government of Mozambique at Maputo. Another Mozambican was made director of conservation. A clinic and school were built to serve the nearest village. For the first time local children were given the opportunity to move up the educational ladder to high school. And beyond. Tonga Torcida, my guide during my first visit to the park in 2010, received a scholarship to attend a college in Tanzania, the first from the Gorongosa area to reach this level of education. Torcida graduated in 2014, returned, and was employed in an executive position at the park.

The big game animal populations of Gorongosa, once the glory of Mozambique's premier national reserve, soon began to regain their prewar strength. Most, like the African elephants, lions, Cape buffalo, hippopotamuses, zebras, and a myriad of antelope species, were allowed to breed up from small stocks of wartime survivors. A few, including hyenas and African wild dogs, were listed to be introduced from stocks in surrounding countries. Nile crocodiles, which are dangerously hard to kill and pull ashore, even by heavily armed hunters, evidently had never dropped in numbers.

In an initiative that pioneers as an example for parks around the world, specialists were invited to census all of Gorongosa's plants and animals. The latter include thousands of invertebrate species ranging in size from nearly invisible springtails to the creatures I personally found most startling: crickets and katydids the size of mice. The collections are housed in new laboratories, and continue to add to future programs of scientific research and education on the park grounds. Planned and led by tropical biologist Piotr Naskrecki, whom I enjoy naming the finest naturalist of my acquaintance in the world, this work, as I write, is moving ahead swiftly. Several colleagues and I have for example identified more than two hundred species of ants, of which 10 percent are new to science.

The Mozambican government has recognized the value of a major park for both tourism and science and encouraged its growth. One of the key steps it has taken was the inclusion of Mount Gorongosa in the official boundary of the park, thereby saving both the seasonal cycle of the Lake Urema floodplain and the water supply of the local subsistence farmers. In the planning phase is aid to improve agriculture around the park perimeter and to help establish councils that protect both the rights of the native residents and the safety of the park wildlife. A great deal has been written about the theory and prospects of this kind of broad-ranging conservation. It has been a pleasure to see such a program in action.

Even in the most favorable of circumstances, a nagging problem in biodiversity restoration is determining its baseline. Natural ecosystems change over time, across millennia, often centuries, and sometimes even decades. The species composing them change genetically, enough in tens or hundreds of millennia to rank as different species. In some kinds of plants, new species can originate in one quick step by the hybridization of two species followed by the doubling of the hybrid's chromosomes, or even just the doubling of chromosomes in one species alone. So, how far back in time should restorers go in establishing the baseline goal of their interventions?

The seeming arbitrariness of the baseline has been used by Anthropocene enthusiasts to accept pauperized floras and faunas as they are, so infiltrated by invasive species as to constitute "novel ecosystems." To lower the bar in such a manner reflects ignorance and unacceptable carelessness. Instead, each puzzle of the baseline can and should be analyzed at the species level, with the composition of the floras and faunas tracked back in time to detect major changes.

The species composition immediately prior to the first major shift that can be ascribed to human activity on the basis of fossil and present-day evidence is the baseline tested and preferred by scientists. For Gorongosa National Park, it is the late Pleistocene prior to invasions by Neolithic people from western Africa. On the U.S. Gulf of Mexico coast, it would be just before either the start of the European incursion or the later clear-cutting of the longleaf pine, the keystone species of the great savanna.

One successful case of baseline choice out of many possible has been the aforementioned restoration of the kelp forests lining the North American Pacific coasts. When the sea otter was driven to near extinction by the fur trade, the sea urchins for which it was the principal predator multiplied prodigiously, consuming kelp until the forests were replaced by "sea-urchin barrens." With the sea otter protected and allowed to multiply to its original population level, the kelp returned, along with the large number of marine species dependent on it. At the opposite extreme, a far more difficult challenge is the restoration of Ireland's primeval forests, the last remnants of which were obliterated centuries ago. Its signature remaining ecosystems are the raised peat bogs.

From a scientist's point of view, the problem of establishing a baseline is not an argument against restoration but a series of fascinating challenges deserving combined research in biodiversity, paleontology, and ecology. This will be one of the challenges met as parks and reserves are made centers of research and education around the world.

Helleborus viridis. Lin.
Elleboro Erba Nocca

Polypodium vulgare Lin.
Polipodie comune

Common polypody fern (Polypodium vulgare) on left, green hellebore
(Helleborus viridis) on right. Gaetano Savi, 1805.

19

HALF-EARTH:
HOW TO SAVE THE BIOSPHERE

A t the end of the day, the central question of biodiversity conservation is how many of the surviving wildlands and the
species within them will be lost before the extinction rate is
returned to the prehuman level. The prehuman rate is now put at one
to ten species extinguished per million species each year. In terms of
a human life span that primordial rate is infinitesimal, essentially zero
in conservation thinking. (Keep in mind also that as many as six million contemporary species remain undiscovered by scientists.) Yet it
also means that the current rate of extinction of the well-known species is up by a multiple of close to one thousand and accelerating—
despite the heroic best efforts of the global conservation movement.

Unstanched hemorrhaging has only one end in all biological systems: death for an organism, extinction for a species. Researchers
who study the trajectory of biodiversity loss are alarmed that within
the century an exponentially rising extinction rate might easily wipe
out most of the species still surviving at the present time.

The crucial factor in the life and death of species is the amount of suitable habitat left to them. The relation between habitat area and number of species has been calculated and refined many times and cited often in scientific and popular literature. It is that a change in area of a habitat, up or down, results in a change in the sustainable number of species by the third to fifth root, most commonly close to the fourth root. In this last case, when, for example, 90 percent of the area is removed, the number that can persist sustainably will descend to about a half. Such is the actual condition of many of the most species-rich localities around the world, including Madagascar, the Mediterranean perimeter, parts of continental southwestern Asia, Polynesia, and many of the islands of the Philippines and the West Indies. If 10 percent of the remaining natural habitat were then also removed—a team of lumbermen might do it in a month—most or all of the surviving resident species would disappear.

If, on the other hand, with the relation of sustainable species to the area of their habitat at the fourth root (the approximate median value), the fraction protected in one-half the global surface is about 85 percent. That fraction can be increased by including within the one-half Earth "hot spots," where the largest numbers of endangered species exist.

Today every sovereign nation in the world has a protected-area system of some kind. All together the reserves number about a hundred sixty-one thousand on land and sixty-five hundred over marine waters. According to the World Database on Protected Areas, a joint project of the United Nations Environmental Program and the International Union for Conservation of Nature, they occupied by 2015 a little less than 15 percent of Earth's land area and 2.8 percent of Earth's ocean area. The coverage is increasing gradually. This trend is encouraging. To have reached the existing level is a tribute to those who have led and participated in the global conservation effort. But is the level enough to not just slow but halt the acceleration of species extinction? Unfortunately, it is in fact nowhere close to enough. Might the upward trend conservation efforts have set be enough during the rest of the century to save most of Earth's biodiversity? That is problematic, but I doubt that it can be, and even then there will be far less biodiversity to save.

Even in the best scenarios of conventional conservation practice the losses should be considered unacceptable by civilized peoples. The declining world of biodiversity cannot be saved by the piece-meal operations in current use alone. It will certainly be mostly lost if conservation continues to be treated as a luxury item in national budgets. The extinction rate our behavior is now imposing on the rest of life, and seems destined to continue, is more correctly viewed as the equivalent of a Chicxulub-sized asteroid strike played out over several human generations.

The only hope for the species still living is a human effort commensurate with the magnitude of the problem. The ongoing mass extinction of species, and with it the extinction of genes and eco-systems, ranks with pandemics, world war, and climate change as among the deadliest threats that humanity has imposed on itself. To those who feel content to let the Anthropocene evolve toward what-ever destiny it mindlessly drifts, I say please take time to reconsider. To those who are steering the growth of reserves worldwide, let me make an earnest request: don't stop, just aim a lot higher.

Populations of species that were dangerously small will have space to grow. Rare and local species previously doomed by devel-opment will escape their fate. The unknown species, apparently at least six million in number, will no longer remain silent and thereby be put at highest risk. People will have closer access to a world that is complex and beautiful beyond our present imagining. We will have more time to put our own house in order for future genera-tions. Living Earth, all of it, can continue to breathe.

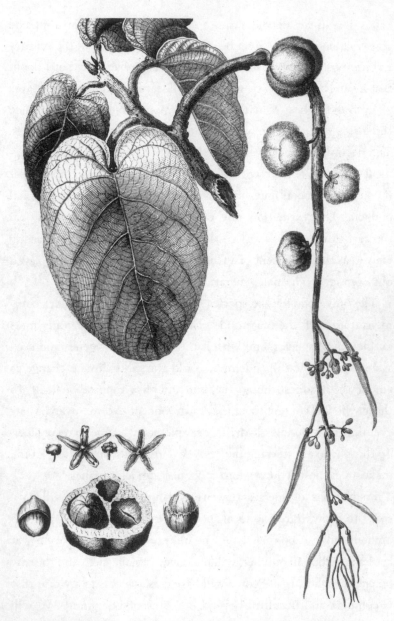

The vine Ronnowia domingensis. *Pierre-Joseph Buc'hoz, 1779.*

20

THREADING THE BOTTLENECK

The Half-Earth solution does not mean dividing the planet into hemispheric halves or any other large pieces the size of continents or nation-states. Nor does it require changing ownership of any of the pieces, but instead only the stipulation that they be allowed to exist unharmed. It does, on the other hand, mean setting aside the largest reserves possible for nature, hence for the millions of other species still alive.

The key to saving one-half of the planet is the ecological footprint, defined as the amount of space required to meet all of the needs of an average person. It comprises the land used for habitation, fresh water, food production and delivery, personal transportation, communication, governance, other public functions, medical support, burial, and entertainment. In the same way the ecological footprint is scattered in pieces around the world, so are Earth's surviving wildlands on the land and in the sea. The pieces range in size from the major desert and forest wildernesses to pockets of restored habitats as small as a few hectares.

But, you may ask, doesn't a rising population and per-capita consumption doom the Half-Earth prospect or any other measure aimed at confining the Anthropocene? It does, but only if the human population continues growing as it has in the past, through the remainder of the twenty-first century and on into the twenty-second century. In this aspect of its biology, however, humanity appears to have won a throw of the demographic dice. Its population growth has begun to decelerate autonomously, without pressure one way or the other from law or custom. In every country where women have gained some degree of social and financial independence, their average fertility has dropped by a corresponding amount through individual personal choice. In Europe and among native-born Americans, it has already reached and continued to hold below the zero-growth threshold of 2.1 children per woman surviving to maturity. Given even a modest amount of personal freedom and an expectation of future security, women choose the option of what ecologists call K-selection, that favoring a small number of healthy well-prepared offspring, as opposed to r-selection, hedging the bet with a larger number of poorly prepared offspring.

There won't be an immediate drop in the total world population. An overshoot still exists due to the longevity of the more numerous per-mother offspring of earlier, more fertile generations. There also remain high-fertility countries, with an average of more than three surviving children born to each woman, thus higher than the 2.1 children per woman that yields zero population growth. They include Patagonia, the Middle East, Pakistan, and Afghanistan, plus all of sub-Saharan Africa exclusive of South Africa. The shift to lower fertility can happen during one or two generations. The United Nations biennial report on population in 2014 projected an

80 percent probability that by 2100 the world population, even as it decelerates toward zero growth, will reach between 9.6 billion and 12.3 billion, up from the 7.2 billion existing in 2014. That is a heavy burden for an already overpopulated planet to bear, but unless women worldwide switch back from the negative population trend of fewer than 2.1 children per woman, a turn downward in the early twenty-second century is inevitable. Another way of looking at the population problem is that it will be solved as an unintended consequence of human nature, namely the flip from r-strategy reproduction to K-strategy reproduction in favorable environments.

But then, what of per-capita consumption? Won't it rise enough to break any resolve for large-scale conservation? It might if the components of the ecological footprints were to remain the same as today. But they will not stay the same. The footprint will evolve, not to claim more and more space, as you might at first suppose, but less. The reason lies in the evolution of the free market system, and the way it is increasingly shaped by high technology. The products that win competition today, and will continue to do so indefinitely, are those that cost less to manufacture and advertise, need less frequent repair and replacement, and give highest performance with a minimum amount of energy. Just as natural selection drives organic evolution by competition among genes to produce more copies of themselves per unit cost in the next generation, raising benefit-to-cost of production drives the evolution of the economy. Almost all of the competition in a free market, other than in military technology, raises the average quality of life. Teleconferencing, online purchase and trade, e-book personal libraries, access on the Internet to all literature and scientific data, online diagnosis and medical practice, food production per hectare sharply raised by indoor vertical gardens with LED lighting, genetically engineered crops and microorganisms, long-distance business conferences and social visits by life-sized images, and not least the best available education in the world free online to anyone, anytime, and anywhere. All of these amenities are now fully available or soon will be. Each will yield more and better results with less per-capita material and energy, and thereby will reduce the size of the ecological footprint.

In viewing the future this way, I wish to suggest a means to achieve almost free enjoyment of the world's best places in the biosphere that I and my fellow naturalists have identified. The cost-benefit ratio would be extremely small. It requires only a thousand or so high-resolution cameras (small and unobtrusive, thanks to the continuing information technology revolution) that broadcast live around the clock from sites within reserves. People would still visit any reserve in the world physically, but they could also travel there virtually and in continuing real time with no more than a few keystrokes in their homes, schools, and lecture halls. Perhaps a Serengeti water hole at dawn? Or the diel cycles of a teeming Amazon canopy? There would also be available streaming video of summer daytime on the coast in the shallow offshore waters of Antarctica, and cameras that continuously travel through the great coral triangle of Indonesia and New Guinea. With species identifications and brief expert commentaries unobtrusively added, the adventure would be forever changing, and safe.

In simplest terms, both shrinkage of the ecological footprint and the resulting improvement of biodiversity conservation are favored because of the acceleration of the replacement of extensive economic growth by intensive economic growth. Extensive economic growth, which prevailed through the twentieth century to the present time, is the increase of per-capita income by adding more capital, more population, and more undeveloped land. Intensive economic growth is that generated by the invention of high-performance new products added to the improved design and use of existing products. The iconic example of the transition is Moore's Law, for its inventor Gordon Moore, cofounder of Intel (and incidentally a leading activist in global conservation): the cost of microchip transistors will fall because the number that can be etched into a fixed area of computer microprocessor will double every two years. The law held between 2002 and 2012, with production rising from 2.6 million per dollar to twenty million per dollar, before beginning to level off.

A closely linked consequence of economic evolution in the twenty-first century is the shift in worldview from wealth based on quantity to wealth based on quality, with the latter made permanent through ecological realism. The central idea is to view the entire planet as an ecosystem, to see Earth as it is and not as we wish it to be. Given that stability in the economy and in the environment are closely linked, they both require striving for the quality of life through self-understanding as opposed to the conventional accumulation of material wealth, based on the assumption that the wealth can eventually be traded for quality of life.

The ecological-realism worldview has been expressed with resonance in *People and the Planet*, a report of Britain's Royal Society. Its recommendations have been endorsed by the global network of national science academies.

In the most developed and the emerging economies unsustainable consumption must be urgently reduced. This will entail scaling back or radical transformation of damaging material consumption and emissions and the adoption of sustainable technologies, and is critical to ensuring a sustainable future for all. At present, consumption is closely linked to economic models based on growth. Improving the wellbeing of individuals so that humanity flourishes rather than survives requires moving from current economic measures to fully valuing natural capital. Decoupling economic activity from material and environmental throughputs is needed urgently.

The pathway of economic evolution will be set by growth that is increasingly intensive and less extensive. Its most advanced products will empower individuals with instruments capable of achieving more and more with less and less per-capita consumption of materials and energy. The environmental consequence of its success will be the opposite of that envisioned by supporters of Anthropocene ideology. Achieved, it will reduce the world ecological footprint, and free space and resources for the rest of life, not constrict it as a perceived necessity of a growing economy. The biosphere and the ten million species that compose it will no longer be treated as a commodity, but as something vastly more important—a mysterious entity still beyond the boundaries of our imagination yet vital to long-term human existence.

With innovation and effort, we will find a way to steer through the climate-change crisis without having to resort to the gargantuan and dangerous programs of geoengineering now being discussed. In particular, let us hope that desperation from fear of a declining future will not force the human race to scrub the atmosphere of excess carbon dioxide, then somehow put the carbon gathered back into the ground. Or, alternatively, coat Earth's surface with sulfates to reflect part of the sun's energy. Or, even worse but still talked about, add lime to seawater to absorb excess carbon dioxide out of the atmosphere.

The spearhead of intensive economic evolution, and with it hope for biodiversity, is contained in the linkage of biology, nanotechnology, and robotics. Two ongoing enterprises within it, the creation of artificial life and artificial minds, seem destined to preoccupy a large part of science and high technology for the rest of the present century. They are also by happenstance well on track to help reduce the ecological footprint, providing a better quality of life with less energy and resources. This result should yield an unintended consequence of entrepreneurial innovation, in this case participating in the protection of Earth's biodiversity for future generations.

The creation of artificial life-forms is already a reality. On May 20, 2010, a team of researchers at the J. Craig Venter Institute in California announced the second genesis of life, this time by human rather than divine command. They had built live cells from the ground up. With simple chemical reagents off the shelf, they assembled the entire genetic code of a bacterial species, *Mycoplasma mycoides*, a double helix of 1.08 million DNA base pairs. During the process they modified the code sequence slightly, implanting a statement made by the late theoretical physicist Richard Feynman, "What I cannot create, I do not understand," in order to detect daughters of the altered mother cells in future tests. Then they transplanted the altered DNA into a recipient cell from which the original DNA had been removed. The newly encoded cell fed and multiplied like a natural cell.

The entity was given a seventeenth century Latinized name with an appropriately robotic surname, *Mycoplasma mycoides* JCVI-syn 1.0. Hamilton O. Smith, speaking for the research team, wrote that with this synthetic entity, and with the new tools and techniques designed to complete the project, "We now have the means to dissect the genetic instruction set of a bacterial cell to see and understand how it really works."

In fact, the new technology is ready to do a great deal more. In 2014, a second team, led by Jef Boeke at Johns Hopkins University, constructed a completely artificial chromosome of a yeast cell. This feat also represents an important advance. Yeast cells are more complex than bacterial cells in possessing organelles such as chromosomes and mitochondria.

The textbook example of elementary artificial selection of the past ten millennia is the transformation of teosinte, a species of wild grass with three races in Mexico and Central America, into maize (corn). The food found in the ancestor was a meager packet of hard kernels. Over centuries of selective breeding it was altered into its modern form. Today maize, after further selection and widespread hybridization of inbred strains that display "hybrid vigor," is the principal food of hundreds of millions.

The first decade of the present century thus saw the beginning of the next new major phase of genetic modification beyond hybridization: artificial selection and even direct substitution in single organisms of one gene for another. If we use the trajectory of progress in molecular biology during the previous half century as a historical guide, it appears inevitable that scientists will begin routinely to build cells of wide variety from the ground up, then induce them to multiply into synthetic tissues, organs, and eventually entire independent organisms of considerable complexity.

If people are to live long and healthy lives in the sustainable Eden of our dreams, and our minds are to break free and dwell in the far more interesting universe of reason triumphant over superstition, it will be through advances in biology. The goal is practicable because scientists, being scientists, live with one uncompromising mandate: Press discovery to the limit. Hand the baton from one to the next, if necessary, but never let the effort die. There has already emerged a term for the manufacture of organisms and parts of organisms: synthetic biology. Its potential benefits, easily visualized as spreading through medicine and agriculture, are limited only by imagination. Synthetic biology will also bring onto center stage the microbe-based increase of food and energy.

The potential power of synthetic biology also leads directly to a vexing question: Can we create a human being? Some enthusiasts believe that in time we ultimately can. If scientists succeeded, even if in just good part, we will have drawn close to the Feynman equation: to construct is to understand. But we will also be forced to solve the ultimate problem of philosophy: What is the meaning of humanity?

A note on history is appropriate at this point. A century ago artificial intelligence (AI) engineers and brain scientists began pursuit of separate goals served by different technologies. The primary purpose of artificial intelligence was and remains the creation of devices that perform physical tasks beyond human capacity. Brain science, in contrast, is more focused. Its central and ultimate goal is whole brain emulation (WBE), the modeling and ultimately the construction of a human-grade mind. Today the two endeavors are converging, and in many ways they already overlap. The technology of artificial intelligence has proven essential to whole brain emulation, while activities observed in the living brain promise advances in artificial intelligence.

The greatest challenge in whole brain emulation is the explanation of consciousness. Neurobiologists almost universally agree that it is a material phenomenon with a cellular physical base. As such it is part of what is called the neuronal workspace, and thus subject to experimentation and mapping. Whole brain emulation is still proceeding in baby steps, but each is longer than the one before. If the current trajectory and pace of research can be sustained, it seems likely that WBE will be attained within the present century. Its culmination will rank as one of the greatest achievements of all time. Exactly what will WBE have accomplished? It will have achieved the construction of artificial self-aware minds, reflective, emotional, eager to learn and grow.

Researchers drawn by this goal or key parts of it are unafraid of what they will find and what might become of it. The most successful scientists are like prospectors exploring an unknown territory. What they mostly care about is making a strike—to be the first to find intellectual gold, silver, or oil. People want it, so lay claim to it, let others worry about the consequences. Later in life they become philosophers, and worry. Meanwhile, they are confident that humanity will ultimately be accompanied by man-made intelligence that knows the meaning of intelligence and can be transferred safely to mobile robots. On the other hand, the people in the general public, influenced by Hollywood scriptwriters, are apprehensive. As citizens of still-violent cultures besotted with religious dogma and superstition, even well-educated people are willing to believe almost anything. In AI and WBE they see a blueprint for possible catastrophe. It is easy to imagine human-grade robots gone berserk and wreaking havoc, avatars (robotic human copies) united in revolt against their human creators, and human minds downloaded into computers that are able (as "transhumans") to dominate those who choose to remain mortal flesh and blood. It helps the myths that they are often featured in technically excellent science-fiction films, for example *2001: A Space Odyssey* (1968), *Star Wars* (1977), *The Terminator* (1984), *I, Robot* (2004), *Avatar* (2009), and *Transcendence* (2014), which are among the most entertaining of this genre ever made, thanks to their epic dramas bolstered by brilliant special effects.

The scientists are certain they know better. In any case, we are on our way to moving the brain sciences to the center of biology and the humanities. The magnitude of artificial intelligence is rising as part of the overall exponential growth of machine computing. Measured in calculations performed per second per thousand dollars of hardware, computer performance has increased since 1960 from one ten-thousandth of a calculation per second (one every three hours) to ten billion calculations per second. All of modern civilization, in every country, both developed and developing, has joined the digital revolution. The effect is not reversible. It will continue to intensify relentlessly, and soon it will reach deeply into the lives of everyone. An example is the impact on the longevity of occupations. Carl Benedikt Frey and Michael A. Osborne, economist and mathematician, respectively, at the University of Oxford, have estimated that at least up to 2030, jobs will be relatively secure for recreational therapists, athletic trainers, dentists, clergy, chemical engineers, firefighters, and editors, but at high risk for second-level machinists, secretaries, real estate agents, accountants, auditors, and telemarketers.

Each passing year sees advances in artificial intelligence and their multitudinous applications—advances that would have been thought distantly futuristic a decade earlier. Robots roll over the surface of Mars. They travel around boulders and up and down slopes while photographing, measuring minutiae of topography, analyzing the chemical composition of soil and rocks, and scrutinizing everything for signs of life. In 2014, SCHAFT, a Japanese-built robot, won the International DARPA Robotics Challenge by navigating doorways and debris, cutting a hole in a wall with a power tool, connecting a fire hose, and driving a small automobile along a twisted path. Advanced computers have recently begun to learn and correct themselves with repeated trials. One programmed with this ability trained itself to recognize images of cats. Another programmed to converse at the level of a boy passed the Turing Test (named after the pioneer computer theorist Alan Turing) when a third of a panel of experts speaking with it for five minutes did not recognize it as a machine.

In 1976, Kenneth I. Appel and Wolfgang Haken revolutionized part of mathematics by using ten billion calculations on an early computer to prove the classic four-color map theorem (proof that no more than four colors are needed to draw any two-dimensional map broken into two-dimensional countries or other pieces), where traditional analytic methods to that time had failed. They justified at least in part the remark by Einstein that "God does not care about our mathematical difficulties. He integrates empirically." In other words, where something is countable, He prefers to count. In some way not yet understood, might the same be true of the hundred billion neurons of the natural human brain?

In the early period of the digital revolution, innovators relied on machine design of computers without reference to the human brain, much as the earliest aeronautical engineers used mechanical principles and intuition to design aircraft instead of imitating the flight of birds. Their approach was required by simple ignorance. Neither computer technologists nor brain scientists were advanced enough to make practical connections to living organisms. With the contemporary swift growth of both fields, analogies and even one-on-one comparisons of processes in natural versus man-made are multiplying. The alliance of computer technology and brain science has given birth to whole brain emulation as one of the ultimate goals of science.

Do brain scientists know enough of the brain's circuits and processes to translate them into the algorithms of artificial intelligence? The orientations of the two disciplines remain different. Whereas artificial intelligence is substantially an engineering enterprise, seeking solutions to problems, whole brain emulation is focused on the central problem of brain and mind. The two are nevertheless thoroughly intertwined. According to Daniel Eth and his coworkers at Stanford University, it is realistic to conceive of simulating a human brain in its entirety on a computer, including its thoughts, feelings, memories, and skills. They identify four requisite technologies: first scanning brain cellular architecture totally, next translating the scan into a model, then running the model on a computer, and finally simulating sensory input from the body and the surrounding environment. All this, they and many others believe, can be accomplished well before the end of this century.

For their part, "neuromorphic engineers," researchers centered on computer development, foresee success in developing computers with characteristics that brains have and computers still lack. According to Karlheinz Meier of the University of Heidelberg, three big problems must be solved to use this reverse engineering successfully. The first is that supercomputers currently trying to emulate brains require millions of watts of power, where human brains use only about twenty watts. Another obstacle is that computers cannot as yet tolerate even small failures. The loss of just one transistor can wreck a microprocessor, whereas brains handle a constant loss of neurons. Finally, brains learn and change spontaneously during experiences in the environment and the vast complexity of child development, but computers must follow the fixed paths and branches of predetermined algorithms.

In fact, the difficulties facing the architects of whole brain emulation go much deeper than just the traditional obstacles implicit in engineering design. The most obvious is that the human brain is not a product of engineering but of evolution. It is a jury-rigged product, built from what was available at each interval from past evolution, fitted by natural selection to the environment of the moment. Across the 450 million years of vertebrate evolution, and further back into our invertebrate ancestry, the brain has evolved not so much as an organ of thought as an organ of survival. From the beginning it was programmed to run the autonomic controls of respiration and heartbeat, along with the sensory and motor controls of reflex. From the very start as well, the brain housed the staging centers of instinct. It was in these centers that appropriate stimuli (the "sign stimuli" of the ethologists) triggered inborn instincts ("consummatory acts").

From the time of the ancient human-destined line of amphibians, then reptiles, then mammals, the neural pathways of every part of the brain were repeatedly altered by natural selection to adapt the organism to the environment in which it lived. Step-by-step, from the Paleozoic amphibians to the Cenozoic primates, the ancient centers were augmented by newer centers, chiefly in the growing cortex, that added to learning ability. Adaptation to a particular environment by evolving repertories of reflex and instinct expanded to include adaptability to changing environments. All other things being equal, the ability of organisms to function through seasons and across different habitats gave them an edge in the constant struggle to survive and reproduce.

Little wonder, then, that neurobiologists have found the human brain to be densely sprinkled with partially independent centers of unconscious operations, along with all of the operators of rational thought. Located through the cortex in what may look at first like random arrays are the headquarters of process variously for numbers, attention, face recognition, meanings, reading, sounds, fears, values, and error detection. Decisions tend to be made by the brute force of unconscious choice in these centers prior to conscious comprehension. Decision, even for simple physical action, can proceed without awareness. The process was anticipated clearly, and poetically, as far back as 1902 by Henri Poincaré:

> The subliminal self is in no way inferior to the conscious self; it is not purely automatic; it is capable of discernment, it has tact, delicacy; it knows better how to choose, to divine. What do I say? It knows better how to divine than the conscious self, since it succeeds where that has failed. In a word, is not the subliminal self superior to the conscious self?

Next in evolution came consciousness. Brain scientists don't know exactly what it is, but they are getting a grip on the role it plays as the newcomer in the human brain. Stanislas Dehaene, a leading theorist at the Collège de France, picked up Poincaré's thread in 2014 as follows:

> In fact, consciousness supports a number of specific operations that cannot unfold unconsciously. Subliminal information is evanescent, but conscious information is stable—we can hang on to it for as long as we wish. Consciousness also compresses the incoming information, reducing an immense stream of sense data to a small set of carefully selected bite-size symbols. The sampled information can then be routed to another processing stage, allowing us to perform what are fully controlled chains of operations, much like a serial computer. This broadcasting function of consciousness is essential. In humans, it is greatly enhanced by language, which lets us distribute our conscious thoughts across the social network.

What has brain science to do with biodiversity? With the human future coming more closely in focus, including an opening to the fountainhead of intellect, it is time—past time—to explore more carefully our moral reasoning with respect to the rest of life. Human nature evolved along a zigzag path as a continually changing ensemble of genetic traits. For all those millions of years to the beginning in this eleventh hour of the Anthropocene, our species let the biosphere continue to evolve on its own. Then, with scythe and fire, guided more by ignorant instinct than by reason, we changed everything.

The endgame of biodiversity conservation is being played out in the twenty-first century. The explosive growth of digital technology, by transforming every aspect of our lives and changing our self-perception, has made the "bnr" industries (biology, nanotechnology, robotics) the spearhead of the modern economy. These three have the potential either to favor biodiversity or to destroy it. I believe they will favor it, by moving the economy away from fossil fuels to energy sources that are clean and sustainable, by radically improving agriculture with new crop species and ways to grow them, and by reducing the need or even the desire for distant travel. All are primary goals of the digital revolution. Through them the size of the ecological footprint will also be reduced. The average person can expect to enjoy a longer, healthier life of high quality yet with less energy extraction and raw demand put on the land and sea. If we are lucky (and smart), world population will peak at a little more than ten billion people by the end of the century or soon thereafter. Not only will the world population decline afterward but the ecological footprint as well—perhaps precipitously. The reason is that we are thinking organisms trying to understand how the world works. We will come awake.

Meanwhile, digital technologies have serendipitously made it possible to complete the census of global biodiversity, and with it to determine the status of each of the millions of species that compose Earth's fauna and flora. That process is already under way, albeit still far too slowly—with the end in sight in the twenty-third century. We and the rest of life with us are in the middle of a bottleneck of rising population, shrinking resources, and disappearing species. As its stewards we need to think of our species as being in a race to save the living environment. The logical primary goal is to make it through the bottleneck to a better, less perilous existence while carrying through as much of the rest of life as possible. If global biodiversity is given space and security, most of the large fraction of species now endangered will regain sustainability on their own.

Furthermore, advances made in synthetic biology, artificial intelligence, whole brain emulation, and other, similar mathematically based disciplines can be imported to create an authentic, predictive science of ecology. In it the interrelations of species will be explored as fervently as we now search through our own bodies for health and longevity. It is often said that the human brain is the most complex system known to us in the universe. That is incorrect. The most complex is the individual natural ecosystem, and the collectivity of ecosystems comprising Earth's species-level biodiversity. Each species of plant, animal, fungus, and microorganism is guided by sophisticated decision devices. Each is intricately programmed in its own way to pass with precision through its respective life cycle. It is instructed on when to grow, when to mate, when to disperse, and when to shy away from enemies. Even the single-celled *Escherichia coli*, living in the bacterial paradise of our intestines, moves toward food and away from toxins by spinning its tail cilium one way, then the other way, in response to chemosensory molecules within its microscopic body.

How all these minds and the decision-making devices within and around them evolve, and how they interact with ecosystems— whether tightly or loosely—is a vast area of biology that remains mostly uncharted—and still even undreamed by those scientists who devote their lives to it. The analytic techniques coming to bear on neuroscience, on Big Data theory, on interoperability studies, on simulations with robot avatars, and on other comparable enterprises will find applications in biodiversity studies. They are ecology's sister disciplines.

It is past time to broaden the discussion of the human future and connect it to the rest of life. The Silicon Valley dreamers of a digitized humanity have not done that, not yet. They have failed to give much thought at all to the biosphere. With the human condition changing so swiftly, we are losing or degrading to uselessness ever more quickly the millions of species that have run the world independently of us and free of cost. If humanity continues its suicidal ways to change the global climate, eliminate ecosystems, and exhaust Earth's natural resources, our species will very soon find itself forced into making a choice, this time engaging the conscious part of our brain. It is as follows: Shall we be existential conservatives, keeping our genetically based human nature while tapering off the activities inimical to ourselves and the rest of the biosphere? Or shall we use our new technology to accommodate the changes important solely to our own species, while letting the rest of life slip away? We have only a short time to decide.

The Mexican box turtle Cistudo mexicana. Proceedings of
the Zoological Society of London, *1848–1860.*

WHAT MUST BE DONE

In a world gaining so swiftly in biotechnology and rational capability, it is entirely reasonable to envision a global network of inviolable reserves that cover half the surface of Earth. But will people and their self-centered political leaders share with others what is theirs to dispose? It is popular to say that the first rule of altruism is never to expect anyone to do anything contrary to his personal interest. Because of the way the brain has evolved, it is especially difficult to protect distant environments if people imagine even a small amount of personal risk or sacrifice. True altruism, this logic continues, is limited to family, tribe, race, or nation, where our genes are seen as indirectly rewarded by our individual devotion to others. God is thought to favor the creation story of the believer's religion over those of all other religions. A patriot considers the moral precepts of his society to be the best in the world. It is the national anthem, not a hymn to human achievement, that is played for Olympic winners.

Even though self-rewarding behavior dominates human behavior, it does not act alone. An instinct for true altruism exists. It enters the decision centers of the brain most readily when an individual has some degree of power and hence the responsibility to achieve altruistic goals. The driving evolutionary force that has created it is group selection, as distinguished from individual selection. The essence of the process is the following: *if altruism toward other members of the group contributes to the group's success, the benefit the altruist's bloodline and genes received may exceed the loss in genes caused by the individual's altruism.*

Charles Darwin, who originated the broad outline of this idea, having confessed difficulty of grasping it at first (for one thing, he had no concept of genes), nevertheless expressed it clearly in *The Descent of Man*:

> It must not be forgotten that although a high standard of morality gives but a slight or no advantage to each individual man and his children over the other men of the same tribe, yet that an advancement in the standard of morality and an increase in the number of well-endowed men will certainly give an immense advantage to one tribe over another. There can be no doubt that a tribe including many members who, from possessing in a high degree the spirit of patriotism, fidelity, obedience, courage, and sympathy, were always ready to give aid to each other and to sacrifice themselves for the common good, would be victorious over most other tribes; and this would be natural selection. At all times throughout the world tribes have supplanted other tribes; and as morality is one element in their success, the standard of morality and the number of well-endowed men will thus everywhere tend to rise and increase.

This idea of multilevel selection (individual plus group selection), which has since been refined many times, then proved by theory and experiment, can be extended as part of social evolution to unrelated persons inside and out of the same tribe and even to other species. As I and other researchers have argued in earlier work, the phenomenon of biophilia, the innate love of the living process, applies even to the natural world, in other words humanity's ancestral environment.

The living world is in desperate condition. It is suffering steep declines in all the levels of its diversity. It will be helped but not saved by economic measures of its ecological services and potential products. Nor will the perception of God's holy will suffice: traditional religions are pivoted on the salvation of human beings, here and in the afterlife, above all other purposes that can be conceived.

Only a major shift in moral reasoning, with greater commitment given to the rest of life, can meet this greatest challenge of the century. Wildlands are our birthplace. Our civilizations were built from them. Our food and most of our dwellings and vehicles were derived from them. Our gods lived in their midst. Nature in the wildlands is the birthright of everyone on Earth. The millions of species we have allowed to survive there, but continue to threaten, are our phylogenetic kin. Their long-term history is our long-term history. Despite all of our pretenses and fantasies, we always have been and will remain a biological species tied to this particular biological world. Millions of years of evolution are indelibly encoded in our genes. History without the wildlands is no history at all.

We should forever bear in mind that the beautiful world our species inherited took the biosphere 3.8 billion years to build. The intricacy of its species we know only in part, and the way they work together to create a sustainable balance we have only recently begun to grasp. Like it or not, and prepared or not, we are the mind and stewards of the living world. Our own ultimate future depends upon that understanding. We have come a very long way through the barbaric period in which we still live, and now I believe we've learned enough to adopt a transcendent moral precept concerning the rest of life. It is simple and easy to say: Do no further harm to the biosphere.

SOURCES AND FURTHER READING

Prologue

A bibliographic note: I first presented the basic argument for such a massively expanded reserve in *The Future of Life* (2002) and *A Window on Eternity: A Biologist's Walk Through Gorongosa National Park* (2014). The term "Half-Earth" was suggested for this concept by Tony Hiss in the article "Can the World Really Set Aside Half the Planet for Wildlife?" (*Smithsonian* 45(5): 66–78, 2014).

1 The World Ends, Twice

Brown, L. 2011. *World on the Edge* (New York: W. W. Norton).

Chivian, D., et al. 2008. Environmental genetics reveals a single-species ecosystem deep within Earth. *Science* 322(5899): 275–278.

Christner, B. C., et al. 2014. A microbial ecosystem beneath the West Antarctic ice sheet. *Nature* 512(7514): 310–317.

Crist, E. 2013. On the poverty of our nomenclature. *Environmental Humanities* 3: 129–147.

Emmott, S. 2013. *Ten Billion* (New York: Random House).

Kolbert, E. 2014. *The Sixth Extinction* (New York: Henry Holt).

Priscu, J. C., and K. P. Hand. 2012. Microbial habitability of icy worlds. *Microbe* 7(4): 167–172.

Weisman, A. 2013. *Countdown: Our Last, Best Hope for a Future on Earth?* (New York: Little, Brown).

Wuerthner, G., E. Crist, and T. Butler, eds. 2015. *Protecting the Wild: Parks and Wilderness, the Foundation for Conservation* (Washington, DC: Island Press).

2 Humanity Needs a Biosphere

Boersma, P. D., S. H. Reichard, and A. N. Van Buren, eds. 2006. *Invasive Species in the Pacific Northwest* (Seattle: University of Washington Press).

Murcia, C., et al. 2014. A critique of the "novel ecosystem" concept. *Trends in Ecology and Evolution* 29(10): 548–553.

Pearson, A. 2008. Who lives in the sea floor? *Nature* 454(7207): 952–953.

Sax, D. F., J. J. Stachowicz, and S. D. Gaines, eds. 2005. *Species Invasions: Insights into Ecology, Evolution, and Biogeography* (Sunderland, MA: Sinauer Associates).

Simberloff, D. 2013. *Invasive species: What Everyone Needs to Know* (New York: Oxford University Press).

Tudge, C. 2000. *The Variety of Life: A Survey and a Celebration of All the Creatures That Have Ever Lived* (New York: Oxford University Press).

White, P. J., R. A. Garrott, and G. E. Plumb. 2013. *Yellowstone's Wildlife in Transition* (Cambridge, MA: Harvard University Press).

Wilson, E. O. 1993. *The Diversity of Life: College Edition* (New York: W. W. Norton)

Wilson, E. O. 2002. *The Future of Life* (New York: Alfred A. Knopf)

Wilson, E. O. 2006. *The Creation: An Appeal to Save Life on Earth* (New York: W. W. Norton).

Womack, A. M., B. J. M. Bohannan, and J. L. Green. 2010. Biodiversity and biogeography of the atmosphere. *Philosophical Transactions of the Royal Society of London B* 365: 3645–3653.

Woodworth, P. 2013. *Our Once and Future Planet: Restoring the World in the Climate Change Century* (Chicago: University of Chicago Press).

3 How Much Biodiversity Survives Today?

Baillie, J. E. M. 2010. *Evolution Lost: Status and Trends of the World's Vertebrates* (London: Zoological Society of London).

Bruns, T. 2006. A kingdom revised. *Nature* 443(7113): 758.

Chapman, R. D. 2009. *Numbers of Living Species in Australia and the World* (Canberra, Australia: Department of the Environment, Water, Heritage, and the Arts).

Magurran, A., and M. Dornelas, eds. 2010. Introduction: Biological diversity in a changing world. *Philosophical Transactions of the Royal Society of London B* 365: 3591–3778.

Pereira, H. M., et al., 2013. Essential biodiversity variables. *Science* 339(6117): 278–279.

Schoss, P. D., and J. Handelsman. 2004. Status of the microbial census. *Microbiology and Molecular Biology Reviews* 68(4): 686–691.

Strain, D. 2011. 8.7 million: A new estimate for all the complex species on Earth. *Science* 333(6046): 1083.

Tudge, C. 2000. *The Variety of Life: A Survey and a Celebration of All the Creatures That Have Ever Lived* (New York: Oxford University Press).

Wilson, E. O. 1993. *The Diversity of Life: College Edition* (New York: W. W. Norton).

Wilson, E. O. 2013. Beware the age of loneliness. *The Economist "The World in 2014,"* p. 143.

4 An Elegy for the Rhinos

Platt, J. R. 2015. How the western black rhino went extinct. *Scientific American Blog Network*, January 17, 2015.

Roth, T. 2004. A rhino named "Emi." *Wildlife Explorer* (Cincinnati Zoo & Botanical Gardens), Sept/Oct: 4-9.

Martin, D. 2014. Ian Player is Dead at 87; helped to save rhinos. *New York Times*, December 5, p. B15.

5 Apocalypses Now

Laurance, W. F. 2013. The race to name Earth's species. *Science* 339(6125): 1275.

Sax, D. F., and S. D. Gaines. 2008. Species invasions and extinction: The future of native biodiversity on islands. *Proceedings of the National Academy of Sciences U.S.A.* 105(suppl. 1): 1490–1497.

6 Are We as Gods?

Brand, S. 1968. "We are as gods and might as well get good at it." In *Whole Earth Catalog* (Published by Stewart Brand).

Brand, S. 2009. "We are as gods and HAVE to get good at it." In *Whole Earth Discipline: An Ecopragmatist Manifesto* (New York: Viking).

7 Why Extinction Is Accelerating

Laurance, W. F. 2013. The race to name Earth's species. *Science* 339(6125): 1275.

Hoffman, M., et al. 2010. The impact of conservation on the status of the world's vertebrates. *Science* 330(6010): 1503–1509.

Sax, D. F., and S. D. Gaines. 2008. Species invasions and extinction: The future of native biodiversity on islands. *Proceedings of the National Academy of Sciences U.S.A.* 105(suppl. 1): 1490–1497.

8 The Impact of Climate Change: Land, Sea, and Air

Banks-Leite, C., et al. 2012. Unraveling the drivers of community dissimilarity and species extinction in fragmented landscapes. *Ecology* 93(12): 2560–2569.

Botkin, D. B., et al. 2007. Forecasting the effects of global warming on biodiversity. *BioScience* 57(3): 227–236.

Burkhead, N. M. 2012. Extinction rates in North American freshwater fishes, 1900–2010. *BioScience* 62(9): 798–808.

Carpenter, K. E., et al. 2008. One-third of reef-building corals face elevated extinction risk from climate change and local impacts. *Science* 321(5888): 560–563.

Cicerone, R. J. 2006. *Finding Climate Change and Being Useful.* Sixth annual John H. Chafee Memorial Lecture (Washington, DC: National Council for Science and the Environment).

Culver, S. J., and P. F. Rawson, eds. 2000. *Biotic Response to Global Change: The Last 145 Million Years* (New York: Cambridge University Press).

De Vos, J. M., et al. 2014. Estimating the normal background rate of species extinction. *Conservation Biology* 29(2): 452–462.

Duncan, R. P., A. G. Boyer, and T. M. Blackburn. 2013. Magnitude and variation of prehistoric bird extinctions in the Pacific. *Proceedings of the National Academy of Sciences U.S.A.* 110(16): 6436–6441.

Dybas, C. L. 2005. Dead zones spreading in world oceans. *BioScience* 55(7): 552–557.

Erwin, D. H. 2008. Extinction as the loss of evolutionary history. *Proceedings of the National Academy of Sciences U.S.A.* 105(suppl. 1): 11520–11527.

Estes, J. A., et al. 2011. Trophic downgrading of planet Earth. *Science* 333(6040): 301–306.

Gillis, J. 2014. 3.6 degrees of uncertainty. *New York Times*, December 16, 2014, p. D3.

Hawks, J. 2012. Longer time scale for human evolution. *Proceedings of the National Academy of Sciences U.S.A.* 109(39): 15531–15532.

Herrero, M., and P. K. Thornton. 2013. Livestock and global change: Emerging issues for sustainable food systems. *Proceedings of the National Academy of Sciences U.S.A.* 110(52): 20878–20881.

Jackson, J. B. C. 2008. Ecological extinction in the brave new ocean. *Proceedings of the National Academy of Sciences U.S.A.* 105(suppl. 1): 11458–11465.

Jeschke, J. M., and D. L. Strayer. 2005. Invasion success of vertebrates in Europe and North America. *Proceedings of the National Academy of Sciences U.S.A.* 102(20): 7198–7202.

Laurance, W. F., et al. 2006. Rapid decay of tree-community composition in Amazonian forest fragments. *Proceedings of the National Academy of Sciences U.S.A.* 103(50): 19010–19014.

LoGuidice, K. 2006. Toward a synthetic view of extinction: A history lesson from a North American rodent. *BioScience* 56(8): 687–693.

Lovejoy, T. E., and L. Hannah, eds. 2005. *Climate Change and Biodiversity* (New Haven, CT: Yale University Press).

Mayhew, P. J., G. B. Jenkins, and T. G. Benton. 2008. A long-term association between global temperature and biodiversity, origination and extinction in the fossil record. *Proceedings of the Royal Society of London B* 275: 47–53.

McCauley, D. J., et al. 2015. Marine defaunation: Animal loss in the global ocean. *Science* 347(6219): 247–254.

Millennium Ecosystems Assessment. 2005. *Ecosystems and Human Well Being, Synthesis.* Summary for Decision Makers, 24 pp.

Pimm, S. L., et al. 2014. The biodiversity of species and their rates of extinction, distribution, and protection. *Science* 344(6187): 1246752-1–10 (doi:10.1126/science.1246752).

Pimm, S. L., and T. Brooks. 2013. Conservation: Forest fragments, facts, and fallacies. *Current Biology* 23: R1098, 4 pp.

Stuart, S. N., et al. 2004. Status and trends of amphibian declines and extinctions worldwide. *Science* 306(5702): 1783–1786.

The Economist. 2014. Deep water. February 22.

Thomas, C. D. 2013. Local diversity stays about the same, regional diversity increases, and global diversity declines. *Proceedings of the National Academy of Sciences U.S.A.* 110(48): 19187–19188.

Urban, M. C. 2015. Accelerating extinction risk from climate change. *Science* 348(6234): 571–573.

Vellend, M., et al. 2013. Global meta-analysis reveals no net change in local-scale plant biodiversity over time. *Proceedings of the National Academy of Sciences U.S.A.* 110(48): 19456–19459.

Wagg, C., et al. 2014. Soil biodiversity and soil community composition determine ecosystem multifunctionality. *Proceedings of the National Academy of Sciences U.S.A.* 111(14): 5266–5270.

9 The Most Dangerous Worldview

Crist, E. 2013. On the poverty of our nomenclature. *Environmental Humanities* 3: 129–147.

Ellis, E. 2009. Stop trying to save the planet. *Wired*, May 6.

Kolata, G. 2013. You're extinct? Scientists have gleam in eye. *New York Times*, March 19.

Kumar, S. 2012. Extinction need not be forever. *Nature* 492(7427): 9.

Marris, E. 2011. *Rambunctious Garden: Saving Nature in a Post-Wild World* (New York: Bloomsbury).

Revkin, A. C. 2012. Peter Kareiva, an inconvenient environmentalist. *New York Times*, April 3.

Rich, N. 2014. The mammoth cometh. *New York Times Magazine*, February 27.

Thomas, C. D. 2013. The Anthropocene could raise biological diversity. *Nature* 502(7469): 7.

Murcia, C., et al. 2014. A critique of the "novel ecosystem" concept. *Trends in Ecology and Evolution* 29(10): 548–553.

Voosen, P. 2012. Myth-busting scientist pushes greens past reliance on "horror stories." *Greenwire*, April 3.

Wuerthner, G., E. Crist, and T. Butler, eds. 2015. *Protecting the Wild: Parks and Wilderness, the Foundation for Conservation* (Washington, DC: Island Press).

Zimmer, C. 2013. Bringing them back to life. *National Geographic* 223(4): 28–33, 35–41.

10 Conservation Science

Balmford, A. 2012. *Wild Hope: On the Front Lines of Conservation Success* (Chicago: University of Chicago Press).

Cadotte, M. C., B. J. Cardinale, and T. H. Oakley. 2008. Evolutionary history predicts the ecological impacts of species extinction. *Proceedings of the National Academy of Sciences U.S.A.* 105(44): 17012–17017.

Discover Life in America (DLIA). 2012. Fifteen Years of Discovery. Report of DLIA, Great Smoky Mountains National Park.

Hoffmann, M., et al. 2010. The impact of conservation on the status of the world's vertebrates. *Science* 330(6010): 1503–1509.

Jeschke, J. M., and D. L. Strayer. 2005. Invasion success of vertebrates in Europe and North America. *Proceedings of the National Academy of Sciences U.S.A.* 102(20): 7198–7202.

Reebs, S. 2005. Report card. *Natural History* 114(5): 14. [The Endangered Species Act of 1973.]

Rodrigues, A. S. L. 2006. Are global conservation efforts successful? *Science* 313(5790): 1051–1052.

Schipper, J., et al. 2008. The status of the world's land and marine mammals: diversity, threat, and knowledge. *Science* 322(5899): 225–230.

Stone, R. 2007. Paradise lost, then regained. *Science* 317(5835): 193.

Taylor, M. F. J., K. F. Suckling, and J. J. Rachlinski. 2005. The effectiveness of the Endangered Species Act: A quantitative analysis. *BioScience* 55(4): 360–367.

11 The Lord God Species

Hoose, P. M. 2004. *The Race to Save the Lord God Bird* (New York: Farrar, Straus and Giroux).

12 The Unknown Webs of Life

Dejean, A., et al. 2010. Arboreal ants use the "Velcro® principle" to capture very large prey. *PLoS One* 5(6): e11331.

Dell, H. 2006. To catch a bee. *Nature* 443(7108): 158.

Hoover, K., et al. 2011. A gene for an extended phenotype. *Science* 333(6048): 1401. [Gypsy moth.]

Hughes, B. B., et al. 2013. Recovery of a top predator mediates negative eutrophic affects on seagrass. *Proceedings of the National Academy of Sciences U.S.A.* 110(38): 15313–15318.

Milius, S. 2005. Proxy vampire: Spider eats blood by catching mosquitoes. *Science News* 168(16): 246.

Montoya, J. M., S. L. Pimm, and R. V. Solé. 2006. Ecological networks and their fragility. *Nature* 442(7100): 259–264.

Moore, P. D. 2005. The roots of stability. *Science* 4437(13): 959–961.

Mora, E., et al. 2011. How many species are there on Earth and in the ocean? *PLoS Biology* 9: e1001127.

Palfrey, J., and U. Gasser. 2012. *Interop: The Promise and Perils of Highly Interconnected Systems* (New York: Basic Books).

Seenivasan, R., et al. 2013. *Picomonas judraskela* gen. et sp. nov.: The first identified member of the Picozoa phylum nov., a widespread group of picoeukaryotes, formerly known as 'picobiliphytes.' *PLoS One* 8(3): e59565.

Ward, D. M. 2006. A macrobiological perspective on microbial species. *Perspective* 1: 269–278.

13 The Wholly Different Aqueous World

Ash, C., J. Foley, and E. Pennisi. 2008. Lost in microbial space. *Science* 320(5879): 1027.

Chang, L., M. Bears, and A. Smith. 2011. Life on the high seas—the bug Darwin never saw. *Antenna* 35(1): 36–42.

Gibbons, S. M., et al. 2013. Evidence for a persistent microbial seed bank throughout the global ocean. *Proceedings of the National Academy of Sciences U.S.A.* 110(12): 4651–4655.

McCauley, D. J., et al. 2015. Marine defaunation: Animal loss in the global ocean. *Science* 347(6219): 247–254.

McKenna, P. 2006. Woods Hole researcher discovers oceans of life. *Boston Globe*, August 7.

Pearson, A. 2008. Who lives in the sea floor? *Nature* 454(7207): 952–953.

Roussel, E. G., et al. 2008. Extending the sub-sea-floor biosphere. *Science* 320(5879): 1046.

14 The Invisible Empire

Ash, C., J. Foley, and E. Pennisi. 2008. Lost in microbial space. *Science* 320(5879): 1027.

Bouman, H. A., et al. 2006. Oceanographic basis of the global surface distribution of *Prochlorococcus* ecotypes. *Science* 312(5775): 918–921.

Burnett, R. M. 2006. More barrels from the viral tree of life. *Proceedings of the National Academy of Sciences U.S.A.* 103(1): 3–4.

Chivian, D., et al. 2008. Environmental genomics reveals a single-species ecosystem deep within Earth. *Science* 322(5899): 275–278.

Christner, B. C., et al. 2014. A microbial ecosystem beneath the West Antarctic ice sheet. *Nature* 512(7514): 310–317.

DeMaere, M. Z., et al. 2013. High level of intergene exchange shapes the evolution of holoarchaea in an isolated Antarctic lake. *Proceedings of the National Academy of Sciences U.S.A.* 110(42): 16939–16944.

Fierer, N., and R. B. Jackson. 2006. The diversity and biogeography of soil bacterial communities. *Proceedings of the National Academy of Sciences U.S.A.* 103(3): 626–631.

Hugoni, M., et al. 2013. Structure of the rare archaeal biosphere and season dynamics of active ecotypes in surface coastal waters. *Proceedings of the National Academy of Sciences U.S.A.* 110(15): 6004–6009.

Johnson, Z. I., et al. 2006. Niche partitioning among *Prochlorococcus* ecotypes along ocean-scale environmental gradients. *Science* 311(5768): 1737–1740.

Milius, S. 2004. Gutless wonder: new symbiosis lets worm feed on whale bones. *Science News* 166(5): 68–69.

Pearson, A. 2008. Who lives in the sea floor? *Nature* 454(7207): 952–953.

Seenivasan, R., et al. 2013. *Picomonas judraskela* gen. et sp. nov.: the first identified member of the Picozoa phylum nov. *PLoS One* 8(3): e59565.

Shaw, J. 2007. The undiscovered planet. *Harvard Magazine* 110(2): 44–53.

Zhao, Y., et al. 2013. Abundant SARⅡ viruses in the ocean. *Nature* 494(7437): 357–360.

15 The Best Places in the Biosphere

The selections described in this chapter are subjective assessments by myself and those chosen at my request by eighteen senior conservation biologists based on extensive field experience. The biologists were: Leeanne Alonso, Stefan Cover, Sylvia Earle, Brian Fisher, Adrian Forsyth, Robert George, Harry Greene, Thomas Lovejoy, Margaret (Meg) Lowman, David Maddison, Bruce Means, Russ Mittermeier, Mark Moffett, Piotr Naskrecki, Stuart Pimm, Ghillean Prance, Peter Raven, and Diana Wall.

16 History Redefined

Tewksbury, J. J., et al. 2014. Natural history's place in science and society. *BioScience* 64(4): 300–310.

Wilson, E. O. 2012. *The Social Conquest of Earth* (New York: W. W. Norton).

Wilson, E. O. 2014. *The Meaning of Human Existence* (New York: W. W. Norton).

17 The Awakening

Andersen, D. 2014. Letter dated August 12, quoted with permission.

Millennium Ecosystems Assessment. 2005. *Ecosystems and Human Well Being, Synthesis*. Summary for Decision Makers, 24 pp.

Running, S. W. 2012. A measurable planetary boundary for the biosphere. *Science* 337(6101): 1458–1459.

18 Restoration

Finch, W., et al. 2012. *Longleaf, Far as the Eye Can See* (Chapel Hill, NC: University of North Carolina Press).

Hiss, T. 2014. Can the world really set aside half the planet for wildlife? *Smithsonian* 45(5): 66–78.

Hughes, B. B., et al. 2013. Recovery of a top predator mediates negative trophic effects on seagrass. *Proceedings of the National Academy of Sciences U.S.A.* 110(38): 15313–15318.

Krajick, K. 2005. Winning the war against island invaders. *Science* 310(5753): 1410–1413.

Tallamy, D. W. 2007. *Bringing Nature Home: How You Can Sustain Wildlife with Native Plants* (Portland, OR: Timber Press).

Wilkinson, T. 2013. *Last Stand: Ted Turner's Quest to Save a Troubled Planet* (Guilford, CT: Lyons Press).

Wilson, E. O. 2014. *A Window on Eternity: A Biologist's Walk Through Gorongosa National Park* (New York: Simon & Schuster).

Woodworth, P. 2013. *Our Once and Future Planet: Restoring the World in the Climate Change Century* (Chicago: University of Chicago Press).

Zimov, S. A. 2005. Pleistocene park: Return of the mammoth's ecosystem. *Science* 308(5723): 796–798.

19 Half-Earth: How to Save the Biosphere

Gunter, M. M., Jr. 2004. *Building the Next Ark: How NGOs Work to Protect Biodiversity* (Lebanon, NH: University Press of New England).

Hiss, T. 2014. Can the world really set aside half the planet for wildlife? *Smithsonian* 45(5): 66–78.

Jenkins, C. N., et al. 2015. US protected lands mismatch biodiversity priorities. *Proceedings of the National Academy of Sciences U.S.A.* 112(16): 5081–5086.

Noss, R. F., A. P. Dobson, R. Baldwin, P. Beier, C. R. Davis, D. A. Dellasala, J. Francis, H. Locke, K. Nowak, R. Lopez, C. Reining, S. C. Trombulak, and G. Tabor. 2011. Bolder thinking for conservation. *Conservation Biology* 26(1): 1–9.

Soulé, M. E., and J. Terborgh, eds. 1999. *Continental Conservation: Scientific Foundations of Regional Networks* (Washington, DC: Island Press).

Steffen, W., et al. 2015. Planetary boundaries: Guiding human development on a changing planet. *Sciencexpress*, January 15, pp. 1–17.

20 Threading the Bottleneck

Aamoth, D. 2014. The Turing test. *Time Magazine*, June 23.

Blewett, J., and R. Cunningham, eds. 2014. *The Post-Growth Project: How the End of Economic Growth Could Bring a Fairer and Happier Society* (London: Green House).

Bourne, J. K., Jr. 2015. *The End of Plenty* (New York: W. W. Norton).

Bradshaw, C. J. A., and B. W. Brook. 2014. Human population reduction is not a quick fix for environmental problems. *Proceedings of the National Academy of Sciences U.S.A.* 111(46): 16610–16615.

Brown, L. R. 2011. *World on Edge: How to Prevent Environmental and Economic Collapse* (New York: W. W. Norton).

Brown, L. R. 2012. *Full Planet, Empty Plates: The New Geopolitics of Food Scarcity* (New York: W. W. Norton).

Callaway, E. 2013. Synthetic biologists and conservationists open talks. *Nature* 496(7445): 281.

Carrington, D. 2014. World population to hit 11 bn in 2100—with 70% chance of continuous rise. *The Guardian*, September 18.

Cohen, J. E. 1995. *How Many People Can the Earth Support?* (New York: W. W. Norton).

Dehaene, S. 2014. *Consciousness and the Brain: Deciphering How the Brain Codes Our Thoughts* (New York: Viking).

Eckersley, P., and A. Sandberg. 2013. Is brain emulation dangerous? *J. Artificial General Intelligence* 4(3): 170–194.

Emmott, S. 2013. *Ten Billion* (New York: Random House).

Eth, D., J.-C. Foust, and B. Whale. 2013. The prospects of whole brain emulation within the next half-century. *J. Artificial General Intelligence* 4(3): 130–152.

Frey, G. B. 2015. The end of economic growth? *Scientific American* 312(1): 12.

Garrett, L. 2013. Biology's brave new world. *Foreign Affairs*, Nov-Dec.

Gerland, P., et al. 2014. World population stabilization unlikely this century. *Science* 346(6206): 234–237.

Graziano, M. S. A. 2013. *Consciousness and the Social Brain* (New York: Oxford University Press).

Grossman, L. 2014. Quantum leap: Inside the tangled quest for the future of computing. *Time*, February 6.

Hopfenberg, R. 2014. An expansion of the demographic transition model: The dynamic link between agricultural productivity and population. *Biodiversity* 15(4): 246–254.

Klein, N. 2014. *This Changes Everything* (New York: Simon & Schuster).

Koene, R., and D. Deca. 2013. Whole brain emulation seeks to implement a mind and its general intelligence through systems identification. *J. Artificial General Intelligence* 4(3): 1–9.

Palfrey, J., and U. Gasser. 2012. *Interop: The Promise and Perils of Highly Interconnected Systems* (New York: Basic Books).

Pauwels, E. 2013. Public understanding of synthetic biology. *BioScience* 63(2): 79–89.

Saunders, D. 2010. *Arrival City: How the Largest Migration in History Is Reshaping Our World* (New York: Pantheon).

Schneider, G. E. 2014. *Brain Structure and Its Origins: In Development and in Evolution of Behavior and the Mind* (Cambridge, MA: MIT Press).

Thackray, A., D. Brock, and R. Jones. 2015. *Moore's Law: The Life of Gordon Moore, Silicon Valley's Quiet Revolutionary* (New York: Basic Books).

The Economist. 2014. The future of jobs. January 18.

The Economist. 2014. DIY chromosomes. March 29.

The Economist. 2014. Rise of the robots. March 29–April 4.

United Nations. 2012. *World Population Prospects* (New York: United Nations).

Venter, J. C. 2013. *Life at the Speed of Light: From the Double Helix to the Dawn of Digital Life* (New York: Viking).

Weisman, A. 2013. *Countdown: Our Last, Best Hope for a Future on Earth?* (New York: Little, Brown).

Wilson, E. O. 2014. *A Window on Eternity: A Biologist's Walk Through Gorongosa National Park* (New York: Simon & Schuster).

Zlotnik, H. 2013. Crowd control. *Nature* 501(7465): 30–31.

21 What Must Be Done

Balmford, A., et al. 2004. The worldwide costs of marine protected areas. *Proceedings of the National Academy of Sciences U.S.A.* 101(26): 9694–9697.

Bradshaw, C. J. A., and B. W. Brook. 2014. Human population reduction is not a quick fix for environmental problems. *Proceedings of the National Academy of Sciences U.S.A.* 111(46): 16610–16615.

Donlan, C. J. 2007. Restoring America's big, wild animals. *Scientific American* 296(6): 72–77.

Hamilton, C. 2015. The risks of climate engineering. *New York Times*, February 12, p. A27.

Hiss, T. 2014. Can the world really set aside half the planet for wildlife? *Smithsonian* 45(5): 66–78.

Jenkins, C. N., et al. 2015. US protected lands mismatch biodiversity priorities. *Proceedings of the National Academy of Sciences U.S.A.* 112(16): 5081–5086.

Mikusiński, G., H. P. Possingham, and M. Blicharska. 2014. Biodiversity priority areas and religions—a global analysis of spatial overlap. *Oryx* 48(1): 17–22.

Pereria, H. M., et al. 2013. Essential biodiversity variables. *Science* 339: 277–278.

Saunders, D. 2010. *Arrival City: How the Largest Migration in History Is Reshaping Our World* (New York: Pantheon).

Selleck, J., ed. 2014. *Biological Diversity: Discovery, Science, and Management*. Special issue of *Park Science* 31(1): 1–123.

Service, R. F. 2011. Will busting dams boost salmon? *Science* 334(6058): 888–892.

Steffen, W., et al. 2015. Planetary boundaries: Guiding human development on a changing planet. *Sciencexpress*, January 15, pp. 1–17.

Stuart, S. N., et al. 2010. The barometer of life. *Science* 328(5975): 177.

Wilson, E. O. 2002. *The Future of Life* (New York: Knopf).

Wilson, E. O. 2014. *A Window on Eternity: A Biologist's Walk Through Gorongosa National Park* (New York: Simon & Schuster).

Wilson, E. O. 2014. *The Meaning of Human Existence* (New York: W. W. Norton).

GLOSSARY

Anthropocene The proposed name for a new geological epoch in which the entire global environment has been altered by humanity.

Anthropocene worldview Applied to nature, the belief that all life should be henceforth valuated primarily or even solely for its importance to human welfare. In its extreme form, the worldview envisions future Earth as entirely enveloped and engineered by humans.

biodiversity The total of variation in organisms, in past times and present, in locations up to and including the entire planet, and organized at three levels: ecosystems, species comprising the ecosystems, and genes prescribing the traits of the species.

biosphere All the organisms alive in the world at any moment, which together form a thin spherical layer around the planet.

ecosystem A locality with particular physical traits and the distinctive species that live within it, such as a lake, a forest patch, a coral reef, a tree, a tree hole, or your mouth and esophagus.

gene The basic unit of heredity, encoded by a particular sequence of DNA units.

genus (*plural*: genera) A group of species, living or extinct, closely related to one another by all having descended from the same ancestral species.

Half-Earth The proposal to set aside for nature half the area of Earth's land and half the area of its seas, in order to halt the accelerating extinction of biodiversity.

species A genetically distinct population or cluster of populations whose members freely interbreed with one another in nature.

APPENDIX

There exist organizations and recent trends in large-scale land and marine conservation that lend credibility to the Half-Earth solution.

The organization already in existence on which a sustainable Half-Earth system can logically be built is the World Heritage Foundation, inaugurated in 1972 and administered today by the United Nations Educational, Scientific and Cultural Organization (UNESCO). The rationale of the World Heritage Foundation is expressed in a United Nations Convention to protect "the world's superb natural and scenic areas and historic sites for the present and future of the entire world citizenry."

Of the 1,007 UN sites listed by 2014, 197 were natural and 31 mixed natural and cultural. Each qualifies in at least one criterion in one or the other of ten categories, the last two of which are fully biological, as follows:

> [The site] IX is an outstanding example representing significant on-going ecological and biological processes in the evolution and development of terrestrial, freshwater, coastal and marine ecosystems, and communities of plants and animals.

> [The site] X contains the most important and significant habitats for in-situ conservation of biological diversity, including those containing threatened species of outstanding universal value from the point of view of science or conservation.

The ambiguity of the final phrase, "*from the point of view of science or conservation,*" can and should be hardened and expanded to include *all* species of an ecosystem. As I've stressed, we don't even know the majority of species on Earth well enough to give them a scientific name, much less discover their place in nature or survival status. Therefore we can't yet assess their role one by one in the future of ecosystems and human life. But we can move decisively with wholesale sweeps. Among those most recent and notable are the following:

- Brazil's minister of the environment signed the legal documents required to fund in perpetuity the Amazon Region Protected Areas program (ARPA), covering 51.2 million hectares, the world's largest network of protected tropical rain forests and three times the size of the entire U.S. National Park System.
- The London-based oil company SOCO International announced that it would abandon its plans to explore for oil in the Virunga National Park, a World Heritage Site of the Democratic Republic of Congo. The park is home to a large amount of biodiversity that includes the critically endangered mountain gorilla, the largest primate in the world.
- Following a celebrity-led campaign in China, the consumption of shark-fin soup plummeted by as much as 70 percent. The Chinese fondness for this delicacy had been ravaging shark populations around the world.
- In the United States, as elsewhere, the building of dams has had a devastating effect on freshwater diversity, being responsible for most of the recorded extinctions of native fish and mollusks. Many of the dams are now being torn down, with the annual rate of removal doubling during the first decade of this century.
- Governments can protect a great deal of the living environment with a single relatively small change in policy. In 2012 USAID announced its first-ever Biodiversity Policy, designed to protect native ecosystems and species through "strategic actions to conserve the world's most important biodiversity, such as stamping out global wildlife trafficking; and a new focus on integrating biodiversity and other development sectors for improved outcomes." This broadening of goals, as I know from personal experience in the field, will offer serious assistance for conservation to the developing nations that most need it.
- The World Parks Congress has conceived of a plan with major potential impact on open-water marine ecosystems. The proposal suggests the creation of large marine protected areas (MPAs) over 20 to 30 percent of the world's seas in which fishing is forbidden. Because fish and other life of open marine

areas disperse constantly, the restored productivity of the MPAs would be shared with adjacent fishing grounds. The average increase in yield in the latter, it was estimated, will yield a million additional jobs, and cost less to monitor and protect than government subsidies now provided to increase yield in the entire, largely unprotected open-water system.

ILLUSTRATION CREDITS

Following are the full citations for the illustrations in the frontispiece and leading each of the twenty-one chapters.

[Frontispiece] Bees, flies, and flowers—Frühlingsbild aus b. Insettenleben in Alfred Edmund Brehm, *63 Chromotafeln aus Brehms Tierleben*, Niedere Tiere, Volumes 7–10 (Leipzig: Bibliographisches Institut, 1883–1884) (Ernst Mayr Library, MCZ, Harvard University).

1 [The World Ends, Twice] Fungi—Plate 27 in Franciscus van Sterbeeck, *Theatrum fungorum oft het Tooneel der Campernoelien* (T'Antwerpen: I. Iacobs, 1675), 19 p.l., 396, [20] p.: front., 36 pl. (26 fold.) port.; 21 cm. (Botany Farlow Library RARE BOOK S838t copy 1 [Plate no. 27 follows p. 244], Harvard University).

2 [Humanity Needs a Biosphere] Swans—Schwarzhalsschwan in Alfred Edmund Brehm, *55 Chromotafeln aus Brehms Tierleben*, Vögel, Volumes 4–6 (Leipzig: Bibliographisches Institut, 1883–1884) (Ernst Mayr Library, MCZ, Harvard University).

3 [How Much Biodiversity Survives Today?] Moth, caterpillar, pupa—Plate IX in Maria Sibylla Merian, *Der Raupen wunderbare Verwandelung und sonderbare Blumen-Nahrung: worinnen durch eine gantz-neue Erfindung der Raupen, Würmer, Sommer-vögelein, Motten, Fliegen, und anderer dergleichen Thierlein Ursprung, Speisen und Veränderungen samt ihrer Zeit* (In Nürnberg: zu finden bey Johann Andreas Graffen, Mahlern; in Frankfurt und Leipzig: bey David Funken,

gedruckt bey Andreas Knortzen, 1679–1683). 2 v. in 1 [4], 102, [8]; [4], 100, [4] p. 100, [2] leaves of plates: ill.; 21 cm. (Plate IX follows p. 16) (Botany Arnold [Cambr.] Ka M54 vol. 2, Harvard University).

4 [An Elegy for the Rhinos] Rhinos— Nashorn in Alfred Edmund Brehm, *52 Chromotafeln aus Brehms Tierleben*, Sängetiere, Volumes 1–3 (Leipzig: Bibliographisches Institute, 1883–1884) (Ernst Mayr Library, MCZ, Harvard University).

5 [Apocalypses Now] Turtles and men—Suppenschildkröte in Alfred Edmund Brehm, *63 Chromotafeln aus Brehms Tierleben*, Niedere Tiere, Volumes 7–10 (Leipzig: Bibliographisches Institute, 1883–1884) (Ernst Mayr Library, MCZ, Harvard University).

6 [Are We as Gods?] Otis—*Otis australis* female Plate XXXVI in *Proceedings of the Zoological Society of London* (Illustrations 1848–1860), 1868, Volume II, Aves, Plates I–LXXVI (Ernst Mayr Library, MCZ, Harvard University).

7 [Why Extinction Is Accelerating] Thylacine—Plate XVIII in *Proceedings of the Zoological Society of London* (Illustrations 1848–1860), Volume I, Mammalia, Plates I–LXXXIII (Ernst Mayr Library, MCZ, Harvard University).

8 [The Impact of Climate Change: Land, Sea, and Air] Starfish—Stachelhäuter in Alfred Edmund Brehm, *63 Chromotafeln aus Brehms Tierleben*, Niedere Tiere, Volumes 7–10 (Leipzig: Bibliographisches Institute, 1883–1884) (Ernst Mayr Library, MCZ, Harvard University).

9 [The Most Dangerous Worldview] Bats—Flugfuchs in Alfred Edmund Brehm, *52 Chromotafeln aus Brehms Tierleben*, Sängetiere, Volumes 1–3 (Leipzig: Bibliographisches Institute, 1883–1884) (Ernst Mayr Library, MCZ, Harvard University).

10 [Conservation Science] Seashells—Plate XXXI in *Proceedings of the Zoological Society of London* (Illustrations 1848–1860), Volume V, Mollusca, Plates I–LI (Ernst Mayr Library, MCZ, Harvard University).

11 [The Lord God Species] Ivory-billed woodpecker and willow oak—Plate 16, M. Catesby, 1729, *The Natural History of Carolina*, Volume I (digital realization of original etchings by Lucie Hey and Nigel Frith, DRPG England; courtesy of the Royal Society©), in *The Curious Mister Catesby: edited for the Catesby Commemorative Trust*, by E. Charles Nelson and David J. Elliott (Athens, GA: University of Georgia Press, 2015).

12 [The Unknown Webs of Life] Snakes—*Thamnocentris [Bothriechis] aurifer* and *Hyla holochlora [Agalychnis moreletii]* Plate XXXII in *Proceedings of the Zoological Society of London* (Illustrations 1848–60), Volume IV, Reptilia et Pisces, Plates I–XXXII et I–XI (Ernst Mayr Library, MCZ, Harvard University).

13 [The Wholly Different Aqueous World] Siphonophore—*Forskalia tholoides* in Ernst Heinrich Philipp August Haeckel, Report on the Siphonophorae collected during the voyage of *H.M.S. Challenger* during 1873–1876. (London:1888) reproduced in *Sociobiology* 1975, Figure 19-2 (Ernst Mayr Library, MCZ, Harvard University).

14 [The Invisible Empire] Beetles—Hirschkäfer in Alfred Edmund Brehm, *63 Chromotafeln aus Brehms Tierleben*, Niedere Tiere, Volumes 7–10 (Leipzig: Bibliographisches Institut, 1883–1884) (Ernst Mayr Library, MCZ, Harvard University).

15 [The Best Places in the Biosphere] Snipe—Waldschnepfe in Alfred Edmund Brehm, *55 Chromotafeln aus Brehms Tierleben*, Vögel, Volumes 4–6 (Leipzig: Bibliographisches Institut, 1883–1884) (Ernst Mayr Library, MCZ, Harvard University).

16 [History Redefined] *Hydrolea crispa* and *Hydrolea dichotoma*—Plate CCXLIV in Hipólito Ruiz et Josepho Pavon, *Flora Peruviana et Chilensis: sive Descriptiones, et icones plantarum Peruvianarum, et Chilensium, secundum systema Linnaeanum digestae, cum characteribus plurium generum evulgatorum reformatis,* auctoribus Hippolyto Ruiz et Josepho Pavon (Madrid: Typis Gabrielis de Sancha, 1798–1802). 3 + v.: ill.; 43 cm. (Botany Gray Herbarium Fol. 2 R85x v. 3, Harvard University).

17 [The Awakening] Fish—*Aploactis milesii* (above) and *Apistes panduratus* (below) in *Proceedings of the Zoological Society of London* (Illustrations 1848–1860), Volume IV, Reptilia et Pisces, Plates I–XXXII et I–XI (Ernst Mayr Library, MCZ, Harvard University).

18 [Restoration] Pine—*Pinus Elliotii* Plate 1 in George Engelmann, *Revision of the genus* Pinus, *and description of* Pinus Elliottii (St. Louis: R. P. Studley & Co., 1880). 29 p. 3 plates. 43 cm. (Botany Arboretum Oversize MH 6 En3, Botany Farlow Library Oversize E57r, Botany Gray Herbarium Fol. 2 En3 [3 copies] copy 2, Harvard University).

19 [Half-Earth: How to Save the Biosphere] *Helleborus viridis* Lin. and *Polypodium vulgare* Lin—Plate XII in Gaetano Savi, *Materia medica vegetabile Toscana* (Firenze: Presso Molini, Landi e Co., 1805), 56 pp., 60 leaves of plates: ill.; 36 cm. (Botany Arnold [Cambr.] Oversize Pd Sa9, Botany Econ. Botany Rare Book DEM 51.2 Savi [ECB folio case 2], Botany Gray Herbarium Fol. 3 Sa9, Harvard University).

20 [Threading the Bottleneck] Vine—*Ronnowia domingensis* Plate IV in Pierre-Joseph Buc'hoz, *Plantes nouvellement découvertes: récemment dénommées et classées,*

représentées en gravures, avec leur descriptions; pour servir d'intelligence a l'histoire générale et économique des trois regnes (Paris: l'Auteur, 1779–1784) (Botany Arnold [Cambr.] Fol. 4 B85.3p 1779, Harvard University).

21 [What Must be Done] Turtle——*Cistudo (Onychotria) mexicana* Gray in *Proceedings of the Zoological Society of London* (Illustrations 1848-60), Volume IV, Reptilia et Pisces, Plates I–XXII et I–XI (Ernst Mayr Library, MCZ, Harvard University).

ACKNOWLEDGMENTS

This essay could not have been completed without the help of many friends and colleagues. I'm especially grateful to my agent John Taylor "Ike" Williams, for both fiscal and legal advice; to Robert Weil, my editor, for inspiration and guidance; to Kathleen M. Horton, for expert research, editing, and manuscript preparation; and to my wife, Renee, for support and advice, as ever before. Gregory C. Carr and MC Davis were essential supporters and coworkers during my visits to Mozambique and Florida. George Keremedjiev, for years a close intellectual companion, introduced me to key trends in the development of high technology and the brain sciences.

I'm further very grateful to my friend Tony Hiss for his encouragement and suggestion that the name "Half-Earth" be applied to my proposal in *The Future of Life*, *A Window on Eternity*, and the present work; and to Paula Ehrlich for her valuable editorial advice.

Those consulted on "The Best Places in the Biosphere," all of who responded, were: Leeanne Alonso, Stefan Cover, Sylvia Earle, Brian Fisher, Adrian Forsyth, Robert George, Harry Greene, Tom Lovejoy, Meg Lowman, David Maddison, Bruce Means, Russ Mittermeier, Mark Moffett, Piotr Naskrecki, Stuart Pimm, Ghillean Prance, Peter Raven, and Diana Wall.

I am indebted to librarians Connie Rinaldo, Mary Sears, and Dana Fisher of the Ernst Mayr Library of the Museum of Comparative Zoology, Harvard University, and to Judith A. Warnement and Lisa DeCesare of the Botanical Library in the Herbarium, Harvard University, for access to and assistance with selecting the figures used in *Half-Earth*.

The butterfly on the cover is the "semialba" form, a seasonal phenotype of the Orange Sulphur (*Colias eurytheme*) of North America (from Rick Cech and Guy Tudor, *Butterflies of the East Coast: An Observer's Guide*, Princeton University Press, 2005).

INDEX

Page numbers in *italics* refer to figures and illustrations.

ABOUT THE AUTHOR

Edward O. Wilson is regarded as one of the world's preeminent biologists and naturalists. The author of more than twenty books, including *The Creation* and the Pulitzer Prize–winning *The Ants* and *Naturalist*, Wilson, a professor emeritus at Harvard University, makes his home in Lexington, Massachusetts.